Disclaimer

The FEEDS Directory is an introductory guide to feedstuffs and as such is not an alternative to nutritional advice. We recommend that a suppliers analysis is used and that all rations are formulated by an experienced nutritionist.

Context accepts no responsibility for errors or omissions in this guide. If you have corrections or suggestions to improve The FEEDS Directory we would welcome your comments.

First published 1997

British Library Cataloguing in Publication Data

The FEEDS Directory - Vol. 1 Commodity Products

I. Ewing, W.N.

ISBN 1-899043-01-2

© Context 1997

Produced and Published by Context
Design: Ian Robinson
Copy Editing: Judith Bennett
Production: Joe Whatnall
Marketing: Llynda Baugh
Distribution: Paula Perkin

Context Publications
45 Swepstone Road
Heather
Leicestershire
England
LE67 2RE
Tel: 01530 264240
Fax: 01530 261579
Email: contextuk@compuserve.com

Also at

CONTEXT
3 Marlborough Park Central
Belfast
Northern Ireland
BT9 1HN

CONTEXT
117 Carrycastle Road
Dungannon
Northern Ireland
BT70 1LT

Categories

Colour coded sections

Cereals and By-Products

Forages and Stock Feeds

Legumes and By-Products

Oilseeds and By-Products

Roots, Fruits and By-Products

Miscellaneous

Icons to depict suitability

 Sheep

 Cattle

 Pigs

 Poultry

 Ideal for feeding straight on farm

CONTEXT

Product Analysis

The analysis given are a best estimate for the commodity products listed. Large variations occur in product analysis especially in forages. Always get a suppliers analysis.

> **NOTE: ALL Analysis quoted are on a Dry Matter basis.**

Abbreviation	Full Name	Units
DM	Dry Matter	%
MER	Metabolisable Energy Ruminants	MJ/kg
MEP	Metabolisable Energy Poultry	MJ/kg
DE	Digestible Energy Pigs	MJ/kg
Oil EE	Oil - Ether Extract	%
Oil AH	Oil - Acid Hydrolysis	%
EFA	Essential Fatty Acids	%
NCGD	Neutral Cellulase Gamanase Digestibility	%
NDF	Neutral Detergent Fibre	%
ASH	Ash	%
FME	Fermentable Metabolisable Energy	MJ/kg
ERDP @ 2	Effective Rumen Degradable Protein at rumen outflow at 0.02 h/1	%
ERDP @ 5	Effective Rumen Degradable Protein at rumen outflow at 0.05 h/1	%
ERDP @ 8	Effective Rumen Degradable Protein at rumen outflow at 0.08 h/1	%
DUP @ 2	Digestible Undegradable Protein at rumen outflow 0.02 h/1	%
DUP @ 5	Digestible Undegradable Protein at rumen outflow 0.05 h/1	%
DUP @ 8	Digestible Undegradable Protein at rumen outflow 0.08 h/1	%
PDIA	Protein undegraded in the rumen and digestible in the small intestine	%
PDIN	Microbial Protein supply where Nitrogen is limiting	%
PDIE	Microbial Protein supply where Energy is limiting	%
Met DI	Digestible Methionine supply to small intestine	%
Lys DI	Digestible Lysine supply to small intestine	%

Inclusion Rates

Suggested inclusion rates are given. However combinations of like products will limit individual inclusion rates eg. Peas vs Beans. When formulating diets other nutritional interactions may occur. It is therefore essential that the advice of a qualified nutritionist is taken.

Index

| **Miscellaneous** | |

Introduction

Manufactured synthetically to replicate naturally occurring amino-acids, these substances have increased in importance as nutrients to improve animal performance. Developed initially for human consumption, demand from intensive livestock production has caused producers to invest in large scale manufacture to provide amino acids at economic prices for inclusion in animal feed.

Origin/Place of Manufacture

USA, Europe, Japan.

Nutritional Benefit

Commercially fed amino acids for pigs and poultry are usually Lysine and Methionine. The amino acids are used to make up any shortfall between nutrient demand and that supplied by raw materials, with their inclusion rate dependent on the high price of other quality ingredients.

Protected forms of Lysine and Methionine are now available for ruminants which allow large proportions of these amino acids to 'bypass' the rumen, which would otherwise be degraded. This has, reportedly, led to improvements in milk yield and milk protein content.

Colour/Texture

White to brown powders. DL Methionine - white; Lysine HCL - off white/brown. Will vary from supplier to supplier.

Palatability

Not fed at high levels or as a straight, so there are no palatablilty problems.

Limits to Usage (Anti-Nutritional Factors) Cost can be prohibitive.

Concentrate Inclusion % per species As recommended by the supplier.

	Inc %	
Dairy	0.001	Methionine
Pig	0.25	Methionine
Chicken	0.2	Methionine
	0.1	Lysine

Storage

Store in dry, clean conditions away from strong light, in sealed packaging from manufacturer.

Alternative Names Many different trade names apply.

Bulk Density 600 - 700 Kg/m³

Typical Analysis	DL-Methionine	L-Lysine	L-Threonine	Typical Analysis	DL-Methionine	L-Lysine	L-Threonine
Dry Matter	99.0	99.0	99.0	DUP (@ 8)	0	0	0
Crude Protein	52.0	93.0	72.0	Salt	0	0	0
DCP	0	0	0	Ca	0	0	0
MER	20.0	16.0	-	Total Phos	0	0	0
MEP	24.0	20.0	14.5	Av Phos	0	0	0
Crude Fibre	0	0	0	Magnesium	0	0	0
Oil (EE)	0	0	0	Potassium	0	0	0
Oil (AH)	0	0	0	Sodium	0	0	0
EFA	0	0	0	Chloride	0	0	0
Ash	0.2	0.2	0.2	Total Lysine	0	79.0	0
NCGD	0	0	0	Avail Lysine	0	-	0
NDF	0	0	0	Methionine	99.0	0	0
ADF	0	0	0	Meth & Cysteine	0	0	0
Starch	0	0	0	Tryptophan	0	0	0
Sugar	0	0	0	Threonine	0	0	98.0
Starch + Sugars	0	0	0	Arginine	0	0	0
FME	0	0	0	PDIA	-	-	-
ERDP (@ 2)	0	0	0	PDIN	-	-	-
ERDP (@ 5)	0	0	0	PDIE	-	-	-
ERDP (@ 8)	0	0	0	Met DI	-	-	-
DUP (@ 2)	0	0	0	Lys DI	-	-	-
DUP (@ 5)	0	0	0				

Starch 0% NDF 0% Other 6.8%

Sugars 0% Ash 0.2%

Protein 93% Oil 0%

CONTEXT

Roots, Fruits and By-Products

Introduction

By-product obtained by pressing apples Malus spp. during the production of apple juice.

Apples are pressed during the winter months with the juice expelled for apple juice or cider. The remaining tissue - skins, pips, stalk (pomace or residue) - is either dried or sold moist. Variable production levels and high moisture content mean it is usually sold into the local market, providing an ideal buy for local farmers.

Origin/Place of Manufacture

Produced in the UK and other temperate climates in Europe.

Nutritional Benefit

A good source of digestible fibre but low in protein. Some products have absorbents (eg. wood shavings) added during manufacture to aid extraction. So nutritional quality will obviously be reduced. As a moist product, it is almost 20% dry matter, low in protein (7.0%), with moderate energy levels. It is highly palatable and ideal for most ruminants as a forage replacer, but needs to be fed with the correct mineral supplementation, as it is naturally a low mineral product.

Colour/Texture

A moist green/brown porridge-friable product.

Palatability

Highly palatable when fresh.

Limits to Usage (Anti-Nutritional Factors)

Variable production levels and consistency make it difficult to rely on in rations. Pectins and pentosans may cause digestive disorders in young ruminants, pigs or young calves/lambs, however, mature ruminants have no problems.

Concentrate Inclusion % per species

	Inc %		Inc %		Inc %
Calf	0	Creep	0	Chick	0
Dairy	20	Weaner	0	Broiler	0
Beef	20	Grower	0	Breeder	0
Lamb	10	Finisher	5	Layer	0
Ewe	5	Sow	5		

Storage/Processing

As a moist feed, it should be clamped and well covered. Apple citrus pulp will store for 6 months.

Alternative Names Apple Citrus.

Bulk Density 150 - 350 Kg/m³

Typical Analysis

Dry Matter	22.0	NCGD	76.0	DUP (@ 5)	1.3	Avail Lysine	-
Crude Protein	7.0	NDF	58.5	DUP (@ 8)	1.7	Methionine	-
DCP	5.0	ADF	26.0	Salt	0.7	Meth & Cysteine	-
MER	9.8	Starch	2.0	Calcium	0.2	Tryptophan	-
MEP	13.5	Sugar	15.0	Total Phos	0.15	Threonine	-
DE	7.5	Starch + Sugars	17.0	Av Phos	0.05	Arginine	-
Crude Fibre	21.0	FME	7.9	Magnesium	0.05	PDIA	-
Oil (EE)	2.8	ERDP (@ 2)	6.3	Potassium	0.5	PDIN	-
Oil (AH)	3.0	ERDP (@ 5)	5.2	Sodium	0.3	PDIE	-
EFA	1.0	ERDP (@ 8)	4.5	Chloride	0.4	Met DI	-
Ash	4.5	DUP (@ 2)	0.8	Total Lysine	-	Lys DI	-

Starch 2% NDF 58.5% Other 10%

Sugars 15% Ash 4.5%

Protein 7% Oil 3%

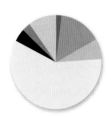

Cereals and By-Products

Introduction

By-product obtained from the manufacture of biscuits, cake or bread.

The waste products are usually a mixture of bread, confectionery, biscuits and even pasta.

These by-products from the food industry are either sub-standard food quality or due to over production. They are usually made from flour (wheat), sugars, flavours and oils, and vary in proportions, depending on the recipe.

Origin/Place of Manufacture

UK and Europe.

Nutritional Benefit

Nutritional values and analysis (especially oil) vary dramatically, depending on products included. Bakery waste has a high energy content due to the high sugar, starch and oil levels, which can be up to 20%. The product may be heat damaged during the cooking process, resulting in reduced starch and protein quality. Usually high in energy which may be derived from soft oils, posing strict limits on its use in ruminant diets.

Colour/Texture

White/cream/brown meal with lumps.

Palatability

Very palatable when fresh. Intakes can be reduced if fed finely ground.

Limits to Usage (Anti-Nutritional Factors)

Beware of high oil products as high inclusion rates can reduce vitamin E in the diet. Care must be taken that plastic packaging, etc. is not included in product.

Avoid biscuit blends which can contain a wide range of by-products other than above.

Concentrate Inclusion % per species

	Inc %		Inc %		Inc %
Calf	7.5	Creep	5	Chick	5
Dairy	20	Weaner	25	Broiler	7.5
Beef	25	Grower	30	Breeder	10.0
Lamb	7.5	Finisher	35	Layer	15
Ewe	15	Sow	30		

Storage/Processing

Does not store well (1-2 weeks), going rancid and mouldy. As the product is hydroscopic, it encourages the uptake of water which further reduces its storage ability.

Alternative Names

By-products from food industry, Biscuit Meal.

Bulk Density 250 kg/m³

Typical Analysis

Dry Matter	88.0	NCDG	70.0	DUP (@ 5)	0.95	Avail Lysine	0.25
Crude Protein	9.5	NDF	7.0	DUP (@ 8)	1.15	Methionine	0.2
DCP	6.1	ADF	6.5	Salt	0.5	Meth & Cysteine	0.35
MER	15.0	Starch	53.5	Ca	0.4	Tryptophan	0.11
MEP	18.5	Sugar	10.0	Total Phos	0.2	Threonine	0.43
DE	18.0	Starch & Sugars	63.5	Av Phos	0.1	Arginine	0.45
Crude Fibre	4.0	FME	12.5	Magnesium	0.2	PDIA	2.6
Oil (EE)	11.0	ERDP (@ 2)	7.9	Potassium	0.25	PDIN	6.9
Oil (AH)	13.0	ERDP (@ 5)	7.3	Sodium	0.4	PDIE	8.9
EFA	4.0	ERDP (@ 8)	6.6	Chloride	0.7	Met DI	0.15
Ash	7.0	DUP (@ 2)	0.8	Total Lysine	0.4	Lys DI	0.39

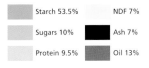

Starch 53.5% NDF 7% Other 0%

Sugars 10% Ash 7%

Protein 9.5% Oil 13%

Cereals and By-Products

Introduction

Grains of Hordeum vulgare L.

Widely grown around the world, with many feed by-products resulting. Barley is usually grown for malting but also grown for animal feed. Top quality products are used in the brewing and distilling industry, with lower quality sold for animal feed. It is an angular grain with a fibrous outer coat.

Origin/Place of Manufacture

Throughout the world, in temperate countries, especially Europe.

Nutritional Benefit

Ideal as a ruminant and non-ruminant feed, with a protein level varying between 6 and 14%, but on average 11-12%. Ruminants benefit from it being high in energy, in the form of starch, making it highly fermentable, encouraging milk protein and fast growth. It is also useful in pig and poultry diets, providing energy from starch. It contains more fibre and less starch than wheat and may be indigestible for young poultry. Nutritional value will depend on the variety, protein level and bushel weight (1000 grain weight). A small proportion (approx 10%) of barley can bypass the rumen unfermented but, as with all cereals, the protein is of average quality being particularly deficient in lysine. Barley is ideal to complement forages but needs careful mineral/vitamin balancing, particularly treated grains. It is especially low in Vitamin A, D, E and calcium. Processing of moist grains with propionic acid reduces the Vitamin E content further.

Colour/Texture
Pale yellow elongated grains.

Palatability
Less palatable than other cereals.

Limits to Usage (Anti-Nutritional Factors)

Ruminants may suffer acidosis (drop in rumen pH) and/or bloat if fed high levels in a feed. Lambs may produce soft fat in their carcass if dietary inclusion is excessive. The presence of beta-glucans cause sticky droppings in poultry and a suitable enzyme should be included (eg. Beta-glucanase). Total feed intakes will reduced if fed finely ground.

Concentrate Inclusion % per species

	Inc %		Inc %		Inc %
Calf	50	Creep	20	Chick	25
Dairy	50	Weaner	25	Broiler	70
Beef	50	Grower	30	Breeder	55
Lamb	25	Finisher	30	Layer	55
Ewe	50	Sow	25		

Storage/Processing

Stores well at moisture below 13%. Normally processed by rolling, grinding, flaking or micronisation which improves the digestibility. Sheep can digest whole barley grain. High inclusion rates will affect compound pellet quality and greater than 70% barley will not pellet easily. It is often treated with propionic acid to preserve it, if high in moisture.

Alternative Names

Bulk Density

Flaked Barley 350 - 390 Kg/m³
Barley Whole 600 - 670 Kg/m³
Barley Meal 400 - 450 Kg/m³

Typical Analysis

Dry Matter	86.0	NCGD	86.0	DUP (@ 5)	1.6	Avail Lysine	0.37
Crude Protein	12.3	NDF	23.1	DUP (@ 8)	1.8	Methionine	0.20
DCP	9.0	ADF	6.4	Salt	0.25	Meth & Cysteine	0.43
MER	13.2	Starch	57.0	Ca	0.1	Tryptophan	0.14
MEP	13.6	Sugar	2.5	Total Phos	0.4	Threonine	0.45
DE	14.5	Starch + Sugars	59.5	Av Phos	0.18	Arginine	0.45
Crude Fibre	5.1	FME	11.0	Magnesium	0.13	PDIA	3.0
Oil (EE)	2.5	ERDP (@ 2)	10.5	Potassium	0.52	PDIN	7.8
Oil (AH)	3.0	ERDP (@ 5)	9.6	Sodium	0.03	PDIE	9.5
EFA	1.1	ERDP (@ 8)	9.3	Chloride	0.16	Met DI	0.18
Ash	2.6	DUP (@ 2)	0.08	Total Lysine	0.42	Lys DI	0.36

Starch 57%   NDF 23.1%   Other 0%

Sugars 2.5%   Ash 2.6%

Protein 12.3%   Oil 2.5%

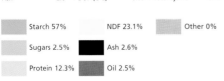

CONTEXT

Cereals and By-Products

Introduction

Grains of Hordeum vulgare L.

This is feed barley treated with sodium hydroxide (caustic) and water to rupture the seed coat and make the whole grain digestible. It is growing in popularity and can be produced on farm in a feeder wagon. It is seen as an alternative to grinding or rolling.

Treatment Process: Add grain to feeder wagon. Add caustic (5%) and mix for 5 minutes. Add water to adjust to 65-70% dry matter. Mix for 10 minutes and leave to stand in a heap for 10 hours (it will heat up). Level out to cool and remix before feeding. Feed after 3-4 days.

Origin/Place of Manufacture

Common in UK, Ireland, Germany, Demark and Sweden.

Nutritional Benefit

Caustic treated barley is claimed to provide all the nutritional benefits of barley but yet be only 30% digestible in the rumen. This whole grain technique is claimed to relocate starch digestion to the intestine to promote more efficient energy utilisation in dairy and beef cows. The sodium bicarbonate coating assists in buffering the rumen.

Colour/Texture

Dark brown/golden. Will darken the longer it is stored. A white 'bloom' often develops naturally on the surface. This is sodium bicarbonate forming during the reaction.

Palatability

Good in a mixture.

Limits to Usage (Anti-Nutritional Factors)

High in sodium, this product needs a low salt mineral. Sodium hydroxide is a hazardous chemical and should be handled with care, adhering to safety instructions of supplier. Storage of moist grains will reduce the Vitamin E level.

Concentrate Inclusion % per species

	Inc %		Inc %		Inc %
Calf	20	Creep	0	Chick	0
Dairy	25	Weaner	0	Broiler	0
Beef	25	Grower	0	Breeder	0
Lamb	0	Finisher	0	Layer	0
Ewe	20	Sow	0		

Storage/Processing

Feed within 4 days of manufacture at 70% dry matter. Adding less water, or salvaging high moisture barley at 80% dry matter allows for long term storage (up to 6 months). Does not need to be ensiled, but needs top and side sheets to keep rain out if stored outside.

Alternative Names Soda Grain, Caustic Barley Grains.

Bulk Density

Typical Analysis

Dry Matter	70.0	NCGD	88.5	DUP (@ 5)	1.6	Av Lysine	-
Crude Protein	12.1	NDF	22.1	DUP (@ 8)	1.8	Methionine	0.20
DCP	9.0	ADF	6.5	Salt	0.25	Meth & Cysteine	0.43
MER	13.0	Starch	55.0	Ca	0.1	Tryptophan	0.14
MEP	-	Sugar	2.5	Total Phos	0.4	Threonine	0.45
DE	-	Starch + Sugars	59.5	Av Phos	-	Arginine	0.45
Fibre	5.0	FME	11.0	Magnesium	0.13	PDIA	2.9
Oil (EE)	2.5	ERDP (@ 2)	10.5	Potassium	0.52	PDIN	3.0
Oil (AH)	3.0	ERDP (@ 5)	9.7	Sodium	3.5	PDIE	9.8
EFA	-	ERDP (@ 8)	9.4	Chloride	0.16	Met DI	0.19
Ash	5.9	DUP (@ 2)	0.08	Total Lysine	0.42	Lys DI	0.35

Starch 55% NDF 22.1% Other 0%
Sugars 2.5% Ash 5.9%
Protein 12.1% Oil 2.5%

Legumes and By-Products

Introduction

Seeds of Vicia faba L. Spp. faba var. equina Pers and var. minuta (Alef) Mansf.

The Vicia Faba or Field Beans are legumes mainly grown for human consumption in Europe. There is a large demand for the food grade product for consumer foods such as baked beans, although varieties are now often grown purely for animal feeding.

Origin/Place of Manufacture

UK and other temperate countries.

Nutritional Benefit

They are an ideal protein source with good levels of energy from starch. High in Lysine but, like the pea, it is low in Methionine and Cysteine, with a low fibre level. Spring beans are higher in protein than winter varieties. They are rich in thiamin and phosphorus. Often interchanged with peas in a ration, although anti-nutritive factors mean they are a second best to peas. Some of the starch (18%) is unfermented in the rumen. They have a high phosphorous content, although its availability may be variable. The oil present is saturated.

Colour/Texture

Mid brown/green large, near oval, but flat seed.

Palatability

Can be unpalatable, but this can be overcome when fed in a mixture.

Limits to Usage (Anti-Nutritional Factors)

They contain tannins, found mainly in the hulls, and may reduce protein digestibility, and some trypsin inhibitors, although new low tannin varieties are grown. Urease, phytates, haemagglutinins and glucosides are regularly present, meaning heat treatment is often necessary. The anti-nutritional factors are removed by processing, and care should be taken to stop mould growth after processing. High inclusion rates may limit performance and careful mineral supplementation is essential.

Concentrate Inclusion % by Species

	Inc %		Inc %		Inc %
Calf	5	Creep	0	Chick	0
Dairy	20	Weaner	0	Broiler	5
Beef	20	Grower	7.5	Breeder	5
Lamb	5	Finisher	10	Layer	5
Ewe	20	Sow	10		

Storage/Processing

They are usually rolled (cracked), coarsely ground or steam/micronised flaked, which improves the starch digestibility. Can go slightly rancid in storage, especially when ground, reducing their palatability. They are usually flaked for younger calves and sheep, coarsely ground for adult cattle and finely ground for non-ruminants. Beans should be dry before processing.

Alternative Names

Field Bean, Horse Bean, Broad Bean.

Bulk Density 590 Kg/m³

Typical Analysis

Dry Matter	86.0	NCDG	92.0	DUP (@ 5)	3.0	Avail Lysine	1.65
Crude Protein	29.0	NDF	21.1	DUP (@ 8)	3.9	Methionine	0.25
DCP	26.0	ADF	12.5	Salt	0.15	Meth & Cysteine	0.6
MER	14.0	Starch	40.0	Ca	0.15	Tryptophan	0.3
MEP	13.5	Sugar	5.5	Total Phos	0.90	Threonine	1.0
DE	15.8	Starch + Sugars	45.5	Av Phos	0.2	Arginine	2.8
Crude Fibre	9.0	FME	12.7	Magnesium	0.2	PDIA	2.8
Oil (EE)	1.8	ERDP (@ 2)	25.0	Potassium	1.2	PDIN	17.6
Oil (AH)	2.2	ERDP (@ 5)	23.5	Sodium	0.05	PDIE	8.9
EFA	0.73	ERDP (@ 8)	22.5	Chloride	0.1	Met DI	0.16
Ash	4.1	DUP (@ 2)	1.5	Total Lysine	1.95	Lys DI	0.65

Starch 38.2% NDF 21% Other 0%

Sugars 5.5% Ash 4.1%

Protein 29% Oil 2.2%

Miscellaneous

Introduction

Product obtained by drying the blood of slaughtered, warm blooded animals.

The product must be substantially free of foreign matter. Dried blood is a product from the slaughtering and meat industry. It is produced at most abbatoirs, although it may be transported to a central drying plant. Fresh blood contains 83% protein in the dry matter but, unfortunately, contains 80% moisture and tterefore needs drying. Ruminant blood meal is no longer used in the U.K.

Origin/Place of Manufacture

UK and around the world.

Nutritional Benefit

Very high in protein with the quality depending on the nature of processing. Excessive heat will reduce the protein digestibility of the product.

Colour/Texture

Red/brown powder.

Palatability

Unpalatable.

Limits to Usage (Anti-Nutritional Factors)

Usage limited by consumer demand and by legislation.

Concentrate Inclusion % per species

	Inc %		Inc %		Inc %
Calf	0	Creep	0	Chick	0
Dairy	0	Weaner	5	Broiler	0
Beef	0	Grower	5	Breeder	0
Lamb	0	Finisher	5	Layer	0
Ewe	0	Sow	0		

Storage/Processing

Bulk Density 490 - 650 kg/m³

Typical Analysis

Dry Matter	90.0	NCGD	-	DUP (@ 5)	-	Avail Lysine	7.5
Crude Protein	83.0	NDF	0	DUP (@ 8)	-	Methionine	1.0
DCP	61.0	ADF	-	Salt	2.0	Meth & Cysteine	2.5
MER	13.2	Starch	0	Ca	0.2	Tryptophan	1.1
MEP	12.7	Sugar	0	Total Phos	0.2	Threonine	3.0
DE	12.8	Starch + Sugars	-	Av Phos	0.1	Arginine	3.9
Fibre	0	FME	-	Magnesium	0.03	PDIA	-
Oil (EE)	1.0	ERDP (@ 2)	-	Potassium	0.15	PDIN	-
Oil (AH)	1.3	ERDP (@ 5)	-	Sodium	0.04	PDIE	-
EFA	-	ERDP (@ 8)	-	Chloride	0.8	Met DI	-
Ash	1.0	DUP (@ 2)	-	Total Lysine	7.0	Lys DI	-

Starch 0% NDF 0% Other 14.5%

Sugars 0% Ash 1%

Protein 83% Oil 1.5%

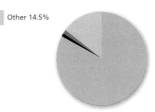

Borage Seed Meal

Oilseeds and By-Products	

Introduction
Borage seed is indigenous to the UK and has historically been regarded as a weed. Grown to produce gamma linoleic acid for use in health and cosmetic products, the seed is crushed to remove its oil (30-35% of the seed).

Origin/Manufacture
UK

Nutritional Benefit
As it is not grown for animal feed, it is usually extremely cheap. High in energy but low in protein digestibility.

Colour/Texture
Blue/grey meal.

Palatability
Poor/average.

Limits to Usage (Anti-Nutritional Factors)
Alkaloids present mean it is unsuitable for young animal diets. Can contain mycotoxins.

Concentrate Inclusion % per species

	Inc %		Inc %		Inc %
Calf	2.5	Creep	0	Chick	0
Dairy	5	Weaner	5.0	Broiler	2.5
Beef	10	Grower	5.0	Breeder	5.0
Lamb	2.5	Finisher	5.0	Layer	5.0
Ewe	10	Sow	5.0		

Storage/Processing
Absorbs moisture and can also go rancid, so avoid storage for long periods.

Alternative Names

Bulk Density

Typical Analysis

Dry Matter	90.0	NCDG	-	DUP (@ 5)	-	Avail Lysine	-
Crude Protein	12.0	NDF	-	DUP (@ 8)	-	Methionine	-
DCP	-	ADF	-	Salt	-	Meth & Cysteine	-
MER	13.6	Starch	-	Ca	0.2	Tryptophan	-
MEP	-	Sugar	-	Total Phos	0.1	Threonine	-
DE	-	Starch + Sugars	-	Av Phos	0.02	Arginine	-
Crude Fibre	8.1	FME	-	Magnesium	-	PDIA	-
Oil (EE)	5.0	ERDP (@ 2)	-	Potassium	-	PDIN	-
Oil (AH)	-	ERDP (@ 5)	-	Sodium	0.02	PDIE	-
EFA	-	ERDP (@ 8)	-	Chloride	-	Met DI	-
Ash	4.5	DUP (@ 2)	-	Total Lysine	-	Lys DI	-

Starch -   NDF -   Other 70.4%

Sugars -   Ash 4.5%

Protein 12%   Oil 5%

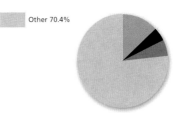

Bread Waste

Miscellaneous

Introduction
Produced either as an oversupply or where bread has passed its retail sell by date.

Origin/Place of Manufacture
UK.

Nutritional Benefit
Good energy levels (14 MJ/Kg in DM), which may be derived from soft oils.
These will reduce its feed value for ruminants. Care should be taken that oils present
have not oxidised.

Colour/Texture
Grey/brown.

Palatability
Good when fresh.

Limits to Usage
As it is a very quickly rumen fermented product, caution should be used with inclusion
rates when fed with finely processed cereals.

Concentrate Inclusion % per species

	Inc %		Inc %		Inc %
Calf	0	Creep	0	Chick	0
Dairy	15	Weaner	15	Broiler	0
Beef	30	Grower	15	Breeder	0
Lamb	0	Finisher	15	Layer	0
Ewe	10	Sow	20		

Storage/Processing
Goes off quickly so should be fed shortly after delivery.

Alternative Names
Crusts

Bulk Density
200 Kg/m³

Typical Analysis

Dry Matter	68.0	NCGD	65.0	DUP (@ 5)	1.6	Avail Lysine	0.38
Crude Protein	12.0	NDF	21.5	DUP (@ 8)	2.1	Methionine	0.25
DCP	11.0	ADF	7.0	Salt	0.4	Meth & Cysteine	0.45
MER	14.0	Starch	26.0	Ca	0.1	Tryptophan	0.18
MEP	16.0	Sugar	7.5	Total Phos	0.4	Threonine	0.35
DE	12.6	Starch + Sugars	33.5	Av Phos	0.2	Arginine	0.4
Crude Fibre	4.5	FME	3.0	Magnesium	0.2	PDIA	-
Oil (EE)	3.0	ERDP (@ 2)	8.9	Potassium	0.1	PDIN	-
Oil (AH)	3.5	ERDP (@ 5)	8.5	Sodium	0.1	PDIE	-
EFA	2.0	ERDP (@ 8)	8.1	Chloride	0.8	Met DI	-
Ash	4.0	DUP (@ 2)	1.2	Total Lysine	0.43	Lys DI	-

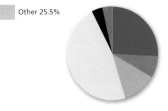

- Starch 26%
- Sugars 7.5%
- Protein 12%
- NDF 21.5%
- Ash 4%
- Oil 3.5%
- Other 25.5%

Cereals and By-Products

Introduction

By-Product of brewing obtained from residues of malted and unmalted cereals and other starchy products.

A moist by-product from the brewing industry made up of spent grains, widely fed to ruminant animals. It can be dried for storage and transportation purposes. Barley, or a variety of other cereals (starchy raw materials can be used), is 'wet mashed'; the liquor is drained for fermentation, and the remaining fibrous and protein material mixed to produce brewers grains. They are commonly used in the dairy industry as a buffer feed or as a forage or concentrate replacer. New beer types, production techniques and alcohol consumption levels have reduced the quantity available. Available throughout the year with the largest supply usually in the summer months.

Origin/Place of Manufacture

Mainly in UK, Ireland and Germany.

Nutritional Benefit

High in digestible fibre and good quality protein which is quite undegradable due to the heat process in manufacture, but low in starch. Ideal for mixing with other forage rations to stimulate dry matter intake and an excellent feed for cattle and sheep. Historically they have been seen as a 'safe feed' providing digestible fibre which can help to buffer more acid feeds. However, today this is less relevant, as modern brewing techniques use small particle grists and the rumen pH in is usually low (4.0-4.2). They are also a good source of phosphorus.

Studies have shown its ability to reduce butterfat content by 0.2-0.3% when compared to grass silage as a sole forage, which may be due to the higher unsaturated oil content of brewers grains. The nutritional value may vary from source to source with dry matters ranging from 18-26%.

Colour/Texture

Pale brown and friable texture.

Palatability

Very palatable to all ruminants and will aid forage intake.

Limits to Usage (Anti-Nutritional Factors)

Brewers grains possess few limiting factors and can be fed to cows at 7-10 kg/head/day, beef cattle ad. lib.and ewes at 3-4 kg/head/day.

Forage Inclusion % per species

	Inc %		Inc %		Inc %
Calf	5	Creep	0	Chick	0
Dairy	30	Weaner	0	Broiler	0
Beef	40	Grower	0	Breeder	0
Lamb	0	Finisher	0	Layer	0
Ewe	5	Sow	0		

Storage/Processing

Either fed shortly after delivery (2-3 weeks) or stored in a sealed clamp/pit.

Alternative Names

Bulk Density

Dry 400 - 480 Kg/m³   **Wet** 880 - 980 Kg/m³

Typical Analysis

Dry Matter	23.0	NCGD	60.0	DUP (@ 5)	5.9	Av Lysine	-
Crude Protein	25.0	NDF	56.5	DUP (@ 8)	7.0	Methionine	0.45
DCP	20.0	ADF	20.0	Salt	0.6	Meth & Cysteine	0.95
MER	11.7	Starch	5.5	Ca	0.4	Tryptophan	0.25
MEP	-	Sugar	1.5	Total Phos	0.5	Threonine	0.8
DE	-	Starch + Sugars	7.0	Av Phos	-	Arginine	1.1
Crude Fibre	17.0	FME	9.0	Magnesium	0.2	PDIA	12.3
Oil (EE)	7.5	ERDP (@ 2)	16.0	Potassium	0.05	PDIN	18.0
Oil (AH)	7.5	ERDP (@ 5)	13.5	Sodium	0.05	PDIE	17.1
EFA	3.0	ERDP (@ 8)	12.5	Chloride	0.2	Met DI	0.4
Ash	4.0	DUP (@ 2)	3.5	Total Lysine	0.84	Lys DI	0.6

Starch 5.5%	NDF 56.5%	Other 0%	
Sugars 1.5%	Ash 4%		
Protein 25%	Oil 7.5%		

Forages and Stock Feeds

Introduction
There are several different types available but the waste is usually made up of stems and leaves after packing brussel sprouts. Only seasonally available in the winter period.

Origin/Manufacture
Throughout the UK.

Nutritional Benefit
Good levels of protein compared to other vegetable wastes but note the high calcium level.

Colour/Texture
Loose green leaves.

Palatability
Palatable to ruminants.

Limits to Usage (Anti-Nutritional Factors)
Contains goitrogenic substances which can interfere with iodine absorption and produce anaemia with prolonged inclusion. Attention to mineral supplementation needed, particularly if fed with other Brassica species eg Rapeseed meal.

Forage Inclusion % per species

	Inc %		Inc %		Inc %
Calf	0	Creep	0	Chick	0
Dairy	15	Weaner	0	Broiler	0
Beef	15	Grower	0	Breeder	0
Lamb	0	Finisher	0	Layer	0
Ewe	15	Sow	0		

Storage/Processing
They should be fed within 2 days of production.

Alternative Names

Bulk Density

Typical Analysis

Dry Matter	15.0	NCGD	-	DUP (@ 5)	-	Meth & Cysteine	-
Crude Protein	24.0	NDF	37.2	DUP (@ 8)	-	Tryptophan	-
DCP	-	ADF	-	Salt	-	Threonine	-
MER	11.5	Starch	4.5	Ca	1.1	Arginine	-
MEP	-	Sugar	20.0	Total Phos	0.3	PDIA	-
DE	-	Starch + Sugars	24.5	Magnesium	-	PDIN	-
Crude Fibre	13.0	FME	-	Potassium	-	PDIE	-
Oil (EE)	2.0	ERDP (@ 2)	-	Sodium	-	Met DI	-
Oil (AH)	2.3	ERDP (@ 5)	-	Chloride	-	Lys DI	-
EFA	-	ERDP (@ 8)	-	Total Lysine	-		
Ash	12.0	DUP (@ 2)	-	Methionine	-		

Starch 4.5%	NDF 37.2%	Other 0%
Sugars 20%	Ash 12%	
Protein 24%	Oil 2.3%	

Forages and Stock Feeds

Introduction

Cabbages can be grown for ruminant animal feed. However more usually they are found in the feed market due to oversupply for human consumption, misshapes, or waste leaves. They have a low proportion of stem to leaf compared to kale and are therefore less fibrous. Usually only seasonally available.

Origin/Place of Manufacture

Eastern UK.

Nutritional Benefit

As a forage supplement.

Colour/Texture

Dark green leaves.

Palatability

Eaten readily by most ruminant animals.

Limits to Usage

Forage Inclusion % per species

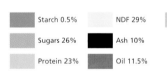

	Inc %		Inc %		Inc %
Calf	0	Creep	0	Chick	0
Dairy	30	Weaner	0	Broiler	0
Beef	30	Grower	0	Breeder	0
Lamb	0	Finisher	0	Layer	0
Ewe	20	Sow	0		

Storage/Processing

Feed immediately after delivery.

Alternative Names

Bulk Density

Typical Analysis

Dry Matter	11.0	NCGD	88.0	DUP (@ 5)	2.0	Av Lysine	-
Crude Protein	23.0	NDF	29.0	DUP (@ 8)	2.7	Methionine	-
DCP	14.0	ADF	25.0	Salt	1.1	Meth & Cysteine	-
MER	11.5	Starch	0.5	Ca	0.8	Tryptophan	-
MEP	-	Sugar	26.0	Total Phos	0.3	Threonine	-
DE	-	Starch + Sugars	26.5	Av Phos	-	Arginine	-
Crude Fibre	12.0	FME	10.0	Magnesium	0.2	PDIA	4.0
Oil (EE)	2.0	ERDP (@ 2)	18.0	Potassium	3.5	PDIN	12.6
Oil (AH)	2.3	ERDP (@ 5)	17.0	Sodium	0.3	PDIE	10.8
EFA	-	ERDP (@ 8)	16.0	Chloride	0.8	Met DI	-
Ash	11.5	DUP (@ 2)	1.0	Total Lysine	-	Lys DI	-

Starch 0.5%	NDF 29%	Other 0%	
Sugars 26%	Ash 10%		
Protein 23%	Oil 11.5%		

Miscellaneous

Introduction

Technically pure magnesium oxide (MgO).

Extracted from mineral quarries and finely ground, this powder is widely used in ruminant rations as a source of magnesium.

Origin/Place of Manufacture

Temperate countries, Turkey, China.

Nutritional Benefit

A good provider of supplemental magnesium.

Colour/Texture

Grey powder.

Palatability

Will make feed unpalatable at high inclusion levels, eg. higher than 2% of concentrate.

Limits to Usage (Anti-Nutritional Factors)

May be contaminated with naturally occurring heavy metals and the necessary guarantees should be requested before using it.

Concentrate Inclusion % per species

	Inc %		Inc %		Inc %
Calf	0	Creep	0	Chick	0
Dairy	1.0 - 5.0	Weaner	0	Broiler	0
Beef	0	Grower	0	Breeder	0
Lamb	0	Finisher	0	Layer	0
Ewe	1.0 - 4.0	Sow	0		

Storage/Processing

Dry storage.

Alternative Names

Cal. Mag., Magnesium Oxide, Mag. ox.

Bulk Density

800 kg/m³

	% added Magnesium in Compound Feed	kg of added Cal Mag in 1 tonne of Compound Feed	Total % Magnesium in Compound Feed
2oz in 16lbs	0.4	8.0	0.6
2oz in 12lb	0.52	10.4	0.72
2oz in 10lb	0.65	13.0	0.85
2oz in 8lb	0.78	15.6	0.98
2oz in 6lb	1.20	24.0	1.40
2oz in 4lb	1.56	31.3	1.76
2oz in 2lb	3.10	62.5	3.30

NB. 2kg Cal. Mag. = 1kg Magnesium

Typical Analysis

Dry Matter	99.5	NDF	0	DUP (@ 8)	0	Methionine	0
Crude Protein	0	ADF	0	Salt	0	Meth & Cysteine	0
MER	0	Starch	0	Ca	0	Tryptophan	0
MEP	0	Sugar	0	Total Phos	0	Threonine	0
DE	0	Starch + Sugars	0	Av Phos	0	Arginine	0
Crude Fibre	0	FME	0	Magnesium	51.0	PDIA	0
Oil (EE)	0	ERDP (@ 2)	0	Potassium	0	PDIN	0
Oil (AH)	0	ERDP (@ 5)	0	Sodium	0	PDIE	0
EFA	0	ERDP (@ 8)	0	Chloride	0	Met DI	0
Ash	100.0	DUP (@ 2)	0	Total Lysine	0	Lys DI	0
NCGD	0	DUP (@ 5)	0	Avail Lysine	0		

Starch 0%		NDF 0%		Other 0%	
Sugars 0%		Ash 100%			
Protein 0%		Oil 0%			

CONTEXT

Carrots

Forages and Stock Feeds			

Introduction

The carrot is a root vegetable grown widely across the UK. Carrots are either surplus or rejected from factories on the basis of shape, size, cuts or even broken carrots. They are available mainly from September to February.

Origin/Manufacture

UK. Available over winter months in local area of processing plant.

Nutritional Benefit

Ideal for ruminants to enhance forage intakes, but contain below average protein levels (9.5%). They have a dry matter of 11-13% and an ME of 12.3-12.8 MJ/Kg DM. They are a good source of beta carotene.

Colour/Texture

Orange, washed, whole/broken roots.

Palatability

Highly palatable.

Limits to Usage (Anti-Nutritional Factors)

As they are high in beta carotene prolonged use at high levels can colour milk fat in dairy cows or carcass fat in beef cattle. Beware of soil contamination.

Forage Inclusion % per species

Typical inclusion 10-15 kg/day for dairy cows and 5kg per 100kg bodyweight for all cattle.

	Inc %		Inc %		Inc %
Calf	0	Creep	0	Chick	0
Dairy	15	Weaner	0	Broiler	0
Beef	20	Grower	0	Breeder	0
Lamb	0	Finisher	0	Layer	0
Ewe	5	Sow	0		

Storage/Processing

They have usually been washed and do not store for more than 14 days in normal conditions.

Alternative Names

Bulk Density

Typical Analysis

Dry Matter	11.0	NDF	20.0	DUP (@ 8)	1.6	Methionine	0.2
Crude Protein	9.5	ADF	15.0	Salt	0.3	Meth & Cysteine	-
DCP	6.0	Starch	10.0	Ca	0.50	Tryptophan	-
MER	12.3	Sugar	30.0	Total Phos	0.35	Threonine	-
MEP	-	Starch + Sugars	40.0	Avail Phos	-	Arginine	-
DE	-	FME	11.9	Magnesium	0.2	PDIA	0.9
Crude Fibre	11.0	ERDP (@ 2)	7.6	Potassium	1.5	PDIN	5.5
Oil (EE)	1.5	ERDP (@ 5)	7.0	Sodium	0.1	PDIE	8.3
Oil (AH)	1.5	ERDP (@ 8)	6.8	Chloride	-	Met DI	0.17
Ash	7.0	DUP (@ 2)	1.0	Total Lysine	0.7	Lys DI	0.62
NCGD	85.0	DUP (@ 5)	1.4	Avail Lysine	-		

- Starch 10%
- Sugars 30%
- Protein 9.5%
- NDF 20%
- Ash 7%
- Oil 1.5%
- Other 22%

Roots, Fruits and By-Products

Introduction

Roots of Manihot esculenta Crantz, regardless of their presentation.

Cassava is a tuberous root of a sub-tropical shrub which is processed before feeding to destroy the cyanide present. It is grown for its starch content and the roots are peeled, chopped and dried after harvesting. The material may come as a meal or pellets, depending on processing method. Its usage depends on price and availability of cereal. Availability may also be affected in Europe by import quotas.

Origin/Manufacture

Tropical and sub-tropical Far East.

Nutritional Benefit

Low in protein and oil but high in starch. The protein is heavily made up of non-protein nitrogen (up to 35%). The analysis will also vary depending on the extent of processing. Ideal for ruminants as starch is slowly degraded.

Colour/Texture

Muddy white meal/pellet or chips.

Palatability

Can vary, depending on cyanide content.

Limits to Usage (Anti-Nutritional Factors)

Linamarin present can release cyanamide which is toxic. Hydrocyanic acid is limited by law and users should consider permitted levels found in the Feeding Stuff Regulations.

Concentrate Inclusion % per species

	Inc %		Inc %		Inc %
Calf	5	Creep	0	Chick	5
Dairy	30	Weaner	10	Broiler	10
Beef	30	Grower	15	Breeder	10
Lamb	5	Finisher	30	Layer	15
Ewe	30	Sow	25		

Storage/Processing

Alternative Names

Manioc, Manihot, Tapioca.

Bulk Density

Typical Analysis

Dry Matter	87.0	NCGD	80.1	DUP (@ 5)	0.45	Avail Lysine	0.05
Crude Protein	3.0	NDF	15.4	DUP (@ 8)	0.63	Methionine	0.5
DCP	1.1	ADF	6.4	Salt	0.2	Meth & Cysteine	0.07
MER	13.2	Starch	71.0	Ca	0.2	Tryptophan	0.03
MEP	14.9	Sugar	3.0	Total Phos	0.2	Threonine	0.07
DE	15.15	Starch + Sugars	74.0	Av Phos	0.15	Arginine	0.15
Crude Fibre	5.0	FME	13.3	Magnesium	0.15	PDIA	0.8
Oil (EE)	0.6	ERDP (@ 2)	2.1	Potassium	1.1	PDIN	1.9
Oil (AH)	1.4	ERDP (@ 5)	1.8	Sodium	0.05	PDIE	8.5
EFA	-	ERDP (@ 8)	1.6	Chloride	0.15	Met DI	0.2
Ash	6.2	DUP (@ 2)	0.16	Total Lysine	0.1	Lys DI	0.05

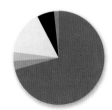

Starch 71%   NDF 15.4%   Other 0%

Sugars 3%   Ash 6.2%

Protein 3%   Oil 1.4%

Cereals and By-Products

Introduction

These are the residues from the processing, storage and shipping of cereal grain and can contain broken or small grains, hulls and other seeds. This is a broad term, as they can come from many grain sources and therefore differ widely in physical nature and nutritive value. They are usually sold separately as barley, wheat, maize or sorghum screening with quality dependent on the time of year and the amount of processing carried out.

Origin/Place of Manufacture

Throughout the world.

Nutritional Benefit

Good quality product but always wise to take a sample for analysis as loads can vary. Ideal for livestock feed and can be close to the original grain in nutritional value when they consist mainly of broken or small grains. However, the fibre content may be increased compared to the original grains.

Colour/Texture

Pale cream, usually a pellet.

Palatability

If dusty, intakes can be reduced.

Limits to Usage (Anti-Nutritional Factors)

Old processed grains may produce a product with moulds present, which will affect performance and even fertility. Inclusion rates for pellets can depend on the particle size of the grain screening. Avoid products with excessive weed seed included.

Concentrate Inclusion % per species

	Inc %		Inc %		Inc %
Calf	10	Creep	0	Chick	0
Dairy	20	Weaner	5	Broiler	0
Beef	30	Grower	15	Breeder	10
Lamb	15	Finisher	25	Layer	15
Ewe	20	Sow	25		

Storage/Processing

Can be very low in bulk density and take up a lot of storage space. Pelleting helps its handling.

Alternative Names

Wheat Screening.

Bulk Density

270 - 370 kg/m³

Typical Analysis

Dry Matter	88.0	NCGD	74.0	DUP (@ 5)	3.1	Avail Lysine	-
Crude Protein	12.2	NDF	40.0	DUP (@ 8)	3.8	Methionine	0.15
DCP	9.1	ADF	-	Salt	0.2	Meth & Cysteine	-
MER	12.0	Starch	34.3	Ca	0.1	Tryptophan	0.2
MEP	9.3	Sugar	2.5	Total Phos	0.15	Threonine	0.4
DE	10.3	Starch + Sugars	36.8	Av Phos	0.06	Arginine	0.7
Crude Fibre	15.0	FME	9.5	Magnesium	0.15	PDIA	-
Oil (EE)	2.0	ERDP (@ 2)	9.5	Potassium	1.2	PDIN	-
Oil (AH)	2.5	ERDP (@ 5)	8.0	Sodium	0.1	PDIE	-
EFA	1.0	ERDP (@ 8)	7.2	Chloride	0.5	Met DI	-
Ash	8.5	DUP (@ 2)	1.8	Total Lysine	0.45	Lys DI	-

Starch 34.3% NDF 40% Other 0%
Sugars 2.5% Ash 8.5%
Protein 12.2% Oil 2.5%

Citrus Pulp Feed

Roots, Fruits and By-Products	

Introduction

By-product obtained by pressing citrus fruits Citrus spp. during the production of citrus juice.

Oranges and other citrus fruits are crushed for their juices (yield 35%+). The remaining solids, consisting of flesh, peel and pips, are either sold moist, locally, or dried for resale, usually in the export market. The dried product is usually ground and pelleted. The inclusion of pips will vary as some manufacturing facilities extract these for their valuable oils.

Origin/Place of Manufacture

America, Southern Europe.

Nutritional Benefit

Similar in digestible fibre levels, energy and fermenting sugar to sugar beet pulp, but slightly lower in protein. The fibre has a buffering capacity in the rumen. Citrus pulp feed can contain oranges, lemons, limes, grapefruit, tangerines, etc. Single source products are the most reliable, and orange citrus feed is recognised as the best as it is more palatable, has a better pH and contains more suitable acids which, in turn, minimise rumen upsets. Analysis may vary if limestone has been added as a drying agent or to stabilise the pH. Ideal for all ruminants and even finisher and sow pig diets and laying poultry. (NB Orange citrus may aid egg colour). The pleasant aroma encourages intake.

Colour/Texture

Brown/orange pellets.

Palatability

Good.

Limits to Usage (Anti-Nutritional Factors)

Product should not be added or removed quickly from a ration as it may cause refusal. Limonin present in the seed is toxic to younger pigs and poultry. Poor calcium to phosphorous ratio.

Concentrate Inclusion % per species

	Inc %		Inc %		Inc %
Calf	20	Creep	0	Chick	0
Dairy	25	Weaner	0	Broiler	2.5
Beef	30	Grower	5	Breeder	5
Lamb	20	Finisher	5	Layer	5
Ewe	30	Sow	10		

Storage/Processing

As a moist product it can be fed immediately or clamped. As a dried product, it will store well for 3 months.

Alternative Names

Orange peel.

Bulk Density

350 - 380 kg/m³

Typical Analysis

Dry Matter	90.0	NDF	37.0	Salt	0.1	Tryptophan	0.25
Crude Protein	6.5	ADF	19.0	Ca	1.6	Threonine	0.1
DCP	3.7	Starch	0.2	Total Phos	0.1	Arginine	0.2
MER	12.2	Sugar	25.5	Av Phos	0.03	TDN	67.0
MEP	6.7	Starch + Sugars	25.7	Magnesium	0.2	PDIA	2.0
DE	12.0	FME	11.6	Potassium	1.2	PDIN	4.4
Crude Fibre	14.0	ERDP (@ 2)	5.1	Sodium	0.05	PDIE	8.9
Oil (EE)	2.6	ERDP (@ 5)	4.6	Chloride	0.05	Met DI	0.1
Oil (AH)	2.9	ERDP (@ 8)	4.1	Total Lysine	0.3	Lys DI	0.3
EFA	0.7	DUP (@ 2)	0.1	Avail Lysine	0.23		
Ash	6.5	DUP (@ 5)	0.5	Methionine	0.1		
NCGD	91.0	DUP (@ 8)	0.8	Meth & Cysteine	0.15		

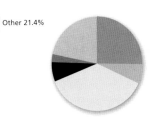

Starch 0.2%	NDF 37%	Other 21.4%
Sugars 25.5%	Ash 6.5%	
Protein 6.5%	Oil 2.9%	

Oilseeds and By-Products

Introduction

By-Product of oil Manufacture, obtained by extraction of dried and roasted cocoa beans, Theabroma cacao L. from which part of the husks has been removed.

A by-product of cocoa manufacture. The beans are removed from large red or yellow pods, roasted, crushed, processed and the inner material ground to make cocoa mass. Further processing to cocoa butter by extrusion or solvent extraction leaves a cocoa residue suitable for animal feed. The extracted cocoa cake is sent for animal feed when in excess supply or of reject quality.

Origin/Place of Manufacture

Beans originate from Africa, especially the western region, and S. America. They are usually processed in the UK or Holland.

Nutritional Benefit

Good protein levels but low in energy. High fibre levels.

Colour/Texture

Coarse brown meal.

Palatability

Some sources have a bitter taste, hence limited inclusion levels recommended.

Limits to Usage (Anti-Nutritional Factors)

Palatability is poor, reducing intake, and fibre levels are high. Theobromine is naturally present. This is a stimulant limited to 700mg/kg in dairy diets, but toxic in pigs, poultry and horses. It is high in potassium and calcium. Avoid cross contamination with other feeds and avoid sporting horse feeds. NB If fed to horses they will fail dope test.

Concentrate Inclusion % per species

	Inc %		Inc %		Inc %
Calf	0	Creep	0	Chick	0
Dairy	2	Weaner	0	Broiler	0
Beef	2	Grower	0	Breeder	0
Lamb	0	Finisher	0	Layer	0
Ewe	2	Sow	0		

Storage/Processing

Alternative Names

Cocoa Bean, extracted.

Bulk Density

475 - 550 kg/m³

Typical Analysis

Dry Matter	90.0	NCGD	-	DUP (@ 5)	-	Av Lysine	-
Crude Protein	18.0	NDF	37.2	DUP (@ 8)	-	Methionine	0.05
DCP	3.8	ADF	-	Salt	0.05	Meth & Cysteine	0.7
MER	4.5	Starch	11.5	Ca	0.24	Tryptophan	0.1
MEP	-	Sugar	2.0	Total Phos	0.4	Threonine	0.5
DE	-	Starch + Sugars	13.5	Av Phos	-	Arginine	0.7
Crude Fibre	16.0	FME	-	Magnesium	0.5	PDIA	-
Oil (EE)	1.0	ERDP (@ 2)	-	Potassium	0	PDIN	-
Oil (AH)	1.3	ERDP (@ 5)	-	Sodium	0.01	PDIE	-
EFA	-	ERDP (@ 8)	-	Chloride	0.03	Met DI	-
Ash	10.0	DUP (@ 2)	-	Total Lysine	0.7	Lys DI	-

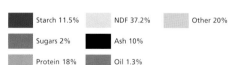

- Starch 11.5%
- Sugars 2%
- Protein 18%
- NDF 37.2%
- Ash 10%
- Oil 1.3%
- Other 20%

Oil seeds and By-Products	

Introduction

By-product of oil manufacture, obtained by pressing the dried kernel (endosperm) and outer husk (tegument) of the seed of the coconut palm.

Copra Expeller is produced from the flesh of dried coconuts after oil expulsion and extraction (60-70% oil), which is used for foods and soaps. The oil is highly saturated, meaning any remaining in the meal is hard enough for animal feeding.

Origin/Place of Manufacture

Equatorial regions, especially Caribbean, Philippines, Kenya, India, S.E. Asia.

Nutritional Benefit

Reasonable levels of digestible fibre make it more suitable for ruminants than non-ruminants. High in protein quality but poor amino acid profile and especially low in lysine and histidine. It is high in fibre (12.5%) reducing the inclusion rate in pigs and poultry. Good Undegradable Protein content. Fat present is saturated and supplies good energy levels.

Colour/Texture

Pale brown meal, pellets or cake.

Palatability

Palatable to ruminants, but poor to others classes of livestock.
Will depend on rancidity/freshness of product.

Limits to Usage (Anti-Nutritional Factors)

Antioxidants should be added and Vitamin E levels monitored.
Oil type may reduce milk fat level. No anti-nutritive factors, but the product should be introduced and removed slowly. Low in methionine and cysteine means careful amino acid supplementation is essential.

Concentrate Inclusion % per species

	Inc %		Inc %		Inc %
Calf	5	Creep	0	Chick	0
Dairy	10	Weaner	0	Broiler	0
Beef	15	Grower	5.0	Breeder	2.5
Lamb	10	Finisher	5.0	Layer	2.5
Ewe	15	Sow	7.5		

Storage/Processing

Care should be taken to avoid product going rancid due to high oil content.
Provides good physical quality to finished product.

Alternative Names

Copra, Coconut Expeller Meal.

Bulk Density

650 - 700 kg/m³

Typical Analysis	Extraction	Expeller	Typical Analysis	Extraction	Expeller
Dry Matter	90.0	91.0	DUP (@ 5)	9.3	9.4
Crude Protein	22.0	23.0	DUP (@ 8)	11.0	11.2
DCP	15.0	15.5	Salt	0.18	0.20
MER	13.0	12.6	Ca	0.3	0.2
MEP	6.2	5.8	Total Phos	0.6	0.53
DE	13.2	12.5	Av Phos	0.2	0.2
Crude Fibre	12.6	12.7	Magnesium	0.3	0.3
Oil (EE)	8.0	2.0	Potassium	2.1	1.2
Oil (AH)	9.0	2.5	Sodium	0.1	0.1
EFA	3.4	1.1	Chlorine	1.2	1.1
Ash	6.8	6.1	Total Lysine	0.56	0.65
NCGD	63.0	66.0	Avail Lysine	0.52	0.54
NDF	45.5	44.2	Methionine	0.35	0.33
ADF	26.9	26.0	Meth& Cysteine	0.65	0.72
Starch	2.0	2.0	Tryptophan	0.19	0.2
Sugar	10.3	9.5	Threonine	0.76	0.78
Starch + Sugars	12.3	11.5	Arginine	2.33	2.73
FME	9.4	9.5	PDIA	11.1	12.2
ERDP (@ 2)	12.2	12.8	PDIN	15.7	16.1
ERDP (@ 5)	9.2	9.3	PDIE	16.5	17.1
ERDP (@ 8)	7.0	7.2	Met DI	0.32	0.35
DUP (@ 2)	5.8	5.6	Lys DI	0.93	1.1

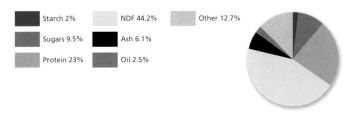

Starch 2% NDF 44.2% Other 12.7%

Sugars 9.5% Ash 6.1%

Protein 23% Oil 2.5%

Miscellaneous

Introduction

Coffee seeds are removed from the outer coating, dried and then roasted. Roasted coffee beans have the coffee extracted into a liquor/syrup for drying to form instant coffee products. Coffee residue remains.

Origin/Place of Manufacture

Processed throughout UK, Europe and USA from beans grown in tropical countries. However, the by-product is usually used in the country of manufacture.

Nutritional Benefit

Very low in energy, high in fibre and of low nutritional quality. It can contain high oil levels, and care should be taken to avoid interference with fibre digestion.

Colour/Texture

Dark brown/black, fine meal.

Palatability

A bitter product which is not usually included at more than 2-4% of the concentrate as intakes will be reduced.

Limits to Usage (Anti-Nutritional Factors)

Can encourage urinary nitrogen and sodium losses and has a strong diuretic effect. Tannins form part of the protein reducing its digestibility, and possibly that of other diet components. Not suitable for horse feeds as it contains caffeine.

Concentrate Inclusion % per species

	Inc %		Inc %		Inc %
Calf	0	Creep	0	Chick	0
Dairy	4	Weaner	0	Broiler	0
Beef	4	Grower	0	Breeder	0
Lamb	0	Finisher	0	Layer	0
Ewe	2	Sow	0		

Storage/Processing

Care should be taken to avoid the oil going rancid.

Alternative Names

Spent Coffee Grounds Meal, Spent Coffee Residue, Dried Coffee Grounds or Cherco.

Bulk Density

500 - 550 kg/m³

Typical Analysis

Dry Matter	91.0	NDF	71.0	Salt	0.05	and Cysteine	0.21
Crude Protein	12.0	ADF	55.0	Ca	0.23	Tryptophan	0.4
DCP	8.5	Starch	3.5	Total Phos	0.07	Threonine	0.3
MER	4.0	Sugar	0.5	Av Phos	0.04	Arginine	0.04
MEP	2.5	Starch + Sugars	4.0	Magnesium	0.03	PDIA	2.5
DE	4.4	FME	5.0	Potassium	0.05	PDIN	7.4
Crude Fibre	44.0	ERDP (@ 2)	6.6	Sodium	0.03	PDIE	7.2
Oil (EE)	2.5	ERDP (@ 5)	5.7	Chloride	0.02	Met DI	0.1
Oil (AH)	2.6	ERDP (@ 8)	5.2	Total Lysine	0.16	Lys DI	0.1
EFA	10.3	DUP (@ 2)	2.2	Avail Lysine	0.10		
Ash	1.7	DUP (@ 5)	3.0	Methionine	0.18		
NCGD	56	DUP (@ 8)	3.6	Methionine			

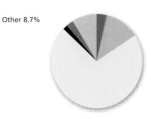

Starch 3.5% NDF 71% Other 8.7%

Sugars 0.5% Ash 1.7%

Protein 12% Oil 2.6%

Cereals and By-Products

Introduction

This is the liquor produced when maize is soaked in the corn wet milling process. It contains valuable protein from the grains and is usually condensed by evaporation.

Origin/Place of Manufacture

UK.

Nutritional Benefit

Contains much of the soluble protein from maize and is ideal in a liquid feed, mixed with molasses, resulting in high energy, protein and sugars. Can improve forage intake as it is very palatable.

Colour/Texture

Yellow/brown viscous liquid.

Palatability

Very palatable/slightly salty.

Limits to Usage (Anti-Nutritional Factors)

Physical consistency may vary and gelling may produce physical problems. pH is low (4-5). Potassium and salt levels are high which can cause scouring at higher levels.

Concentrate Inclusion % per species

	Inc %		Inc %		Inc %
Calf	2	Creep	0	Chick	0
Dairy	10	Weaner	0	Broiler	0
Beef	10	Grower	3	Breeder	0
Lamb	2	Finisher	3	Layer	0
Ewe	3	Sow	3		

Storage/Processing

Not easy to store and may gel. Often mixed with molasses before storage.

Alternative Names

CCSL.

Bulk Density

1200 Kg/m³

Typical Analysis

Dry Matter	45.0	NDF	0	DUP (@ 8)	8.4	Methionine	-
Crude Protein	40.0	ADF	0	Salt	1.0	Meth & Cysteine	-
DCP	32.0	Starch	5.0	Ca	0.4	Tryptophan	-
MER	13.0	Sugar	24.0	Total Phos	3.0	Threonine	-
MEP	-	Starch + Sugars	29.0	Av Phos	2.0	Arginine	-
DE	-	FME	12.3	Magnesium	1.0	PDIA	-
Crude Fibre	0	ERDP (@ 2)	-	Potassium	3.0	PDIN	-
Oil (EE)	1.0	ERDP (@ 5)	-	Sodium	0.8	PDIE	-
Oil (AH)	1.2	ERDP (@ 8)	26.0	Chloride	-	Met DI	-
Ash	17.5	DUP (@ 2)	-	Total Lysine	-	Lys DI	-
NCGD	82.0	DUP (@ 5)	-	Avail Lysine	-		

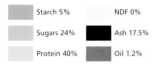

Starch 5% NDF 0% Other 12.3%

Sugars 24% Ash 17.5%

Protein 40% Oil 1.2%

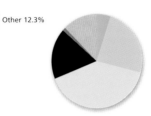

Oilseeds and By-Products

Introduction

By-product of oil manufacture obtained by pressing of seeds from which fibres have been removed.

The seed is grown for its white fibre/cotton for textiles. After by-product of oil manufacture, obtained by pressing of seeds of cotton from which fibres have been removed, the seed is still high in fibre and oil. The oil is removed by expelling or extraction, leaving the cotton seed meal by-product. Some cotton seed meals do not have the fibre coating removed with the oil expelled; this type is high in fibre (20-25%).

Origin/Place of Manufacture

USA, Russia, China, India, Brazil, Argentina. Also processed in Europe.

Nutritional Benefit

A useful mid protein feed which is inferior in quality to soya bean meal as it is higher in fibre and has a poorer amino acid profile (low in lysine, methionine and cystine). It can have a constipating effect. Different sources can vary in quality and the analysis will vary, depending on method of processing, eg. whether it is decorticated or not and whether the oil has been expelled or extracted. Low in Lysine but a useful source of digestible fibre.

Colour/Texture

Brown to yellow, oily, fibrous meal.

Palatability

Low.

Limits to Usage (Anti-Nutritional Factors)

Aflatoxins can be high and guarantees should be sought from suppliers. Gossypol present can produce digestive and palatability upsets, reducing performance at high levels in young animal diets by affecting protein digestibility. Gossypol can affect yolk colour, and is also quite toxic to most monogastrics, and high fibre levels will affect intake.

Both gossypol and aflatoxins need to be within the levels of the Feed Stuff Regulations. Can also be contaminated with salmonella. Not usually fed as a direct raw material on farm.

Concentrate Inclusion % per species

	Inc %		Inc %		Inc %
Calf	5	Creep	0	Chick	0
Dairy	15	Weaner	0	Broiler	0
Beef	20	Grower	0	Breeder	0
Lamb	5	Finisher	2.5	Layer	0
Ewe	10	Sow	5.0		

Storage/Processing

Salmonella may be present, needing treatment. Cotton seed meal may aid physical quality on pelleting.

Alternative Names

Decorticated Cotton Seed Meal.

Bulk Density

550 - 650 kg/m³

Typical Analysis	Extraction	Expellers	Whole
Dry Matter	90.0	92.0	95.0
Crude Protein	41.0	36.0	24.0
DCP	34.0	30.0	18.5
MER	11.4	12.0	15.0
MEP	9.1	9.1	9.7
DE	11.1	13.3	12.0
Crude Fibre	15.9	18.0	27.0
Oil (EE)	1.7	5.5	20.1
Oil (AH)	1.9	6.0	21.1
EFA	1.0	3.4	10.5
Ash	6.0	7.0	6.5
NCGD	75.0	79.0	70.1
NDF	36.0	40.1	42.0
ADF	22.0	25.7	26.3
Starch	3.0	2.0	1.5
Sugar	6.0	6.5	6.5
Starch + Sugars	9.0	8.5	8.0
FME	10.0	9.5	7.5
ERDP (@ 2)	30.1	26.4	15.4
ERDP (@ 5)	25.6	21.9	12.2
ERDP (@ 8)	23.1	19.1	10.7
DUP (@ 2)	7.3	6.1	3.9

Typical Analysis	Extraction	Expellers	Whole
DUP (@ 5)	11.9	9.9	6.6
DUP (@ 8)	14.7	12.3	7.9
Salt	0.1	0.1	0.15
Ca	0.25	0.25	0.25
Total Phos	1.0	1.0	0.75
Av Phos	0.3	0.3	0.3
Magnesium	0.6	0.6	0.4
Potassium	1.5	1.5	1.2
Sodium	0.05	0.05	0.05
Chlorine	0.05	0.05	0.05
Total Lysine	1.8	1.8	1.6
Avail Lysine	1.3	1.3	0.8
Methionine	0.7	0.7	0.6
Meth & Cysteine	1.4	1.3	1.5
Tryptophan	0.6	0.5	0.6
Threonine	1.5	1.2	1.3
Arginine	4.8	4.0	4.7
PDIA	15.4	13.2	5.0
PDIN	29.7	25.5	14.1
PDIE	19.8	17.7	9.0
Met DI	0.28	0.26	0.14
Lys DI	1.1	1.0	0.55

Starch 3%   NDF 36%   Other 6.1%

Sugars 6%   Ash 6%

Protein 41%   Oil 1.9%

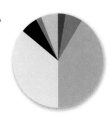

CONTEXT

Miscellaneous	

Introduction

Precipitated calcium monohydrogen phosphate from bones or inorganic sources (CaHPO4 x H20).

Calcium phosphate is usually extracted from rock phosphate by the addition of hydrochloric acid. The mineral is then purified and neutralised to produce Di-calcium Phosphate. Some sources extract the phosphorus from bones.

Origin/Place of Manufacture

Europe, USA, North Africa.

Nutritional Benefit

A major source of supplementary phosphate of reasonable availability. Different sources and manufacturing processes produce phosphorus levels ranging from 17.5% to 21.0%.

Colour/Texture Grey Powder.

Palatability

Not fed straight.

Limits to Usage (Anti-Nutritional Factors)

May be contaminated with traces of fluorine and heavy metals, but guarantees are usually available.

Concentrate Inclusion % per species

	Inc %		Inc %		Inc %
Calf	1	Creep	1	Chick	1
Dairy	1	Weaner	1	Broiler	1
Beef	1	Grower	1	Breeder	1
Lamb	1	Finisher	1	Layer	1
Ewe	1	Sow	1		

Storage/Processing

Stores well as relatively inert.

Alternative Names

DCP, Di-cal.

Bulk Density

640 kg/m³

Typical Analysis

Dry Matter	98.0	NCDG	0	DUP (@ 5)	0	Avail Lysine	0
Crude Protein	0	NDF	0	DUP (@ 8)	0	Methionine	0
DCP	0	ADF	0	Salt	0	Meth & Cysteine	0
MER	0	Starch	0	Ca	23.0	Tryptophan	0
MEP	0	Sugar	0	Total Phos	18.0	Threonine	0
DE	0	Starch + Sugars	0	Av Phos	18.0	Arginine	0
Crude Fibre	0	FME	0	Magnesium	0	PDIA	0
Oil (EE)	0	ERDP (@ 2)	0	Potassium	0	PDIN	0
Oil (AH)	0	ERDP (@ 5)	0	Sodium	0	PDIE	0
EFA	0	ERDP (@ 8)	0	Chlorine	0	Met DI	0
Ash	100.0	DUP (@ 2)	0	Total Lysine	0	Lys DI	0

Starch 0%	NDF 0%	Other 0%	
Sugars 0%	Ash 100%		
Protein 0%	Oil 0%		

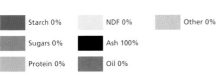

Cereals and By-Products

Introduction

By-product of alcohol distilling obtained by drying solid residues of fermented grain (Hordeum vulgare L.) to which pot ale syrup or evaporated spent wash has been added.

A by-product usually from malt whisky production. Barley is allowed to germinate to produce the malt and is then soaked to release the starch reserves for fermentation. An enzyme is produced in the malting which converts the starch to simpler sugars; yeast is added, fermentation occurs, and the alcohol is distilled off. The grain which remains after the liquor is removed is often called wet draff. This can be pressed, dried, and the spent yeast liquor (Pot Ale Syrup) added back to produce barley dark grain. The product from UK manufacture is usually pelleted.

Origin/Place of Manufacture

Scotland.

Nutritional Benefit

High digestible fibre levels mean these grains are not generally used in pig and poultry rations. As a mid-protein feed, barley distillers are high in Undegraded Protein (UDP) and low in starch due to its extraction. They usually contain copper at approximately 50ppm on a dry matter basis.

Colour/Texture

Dark brown, usually pelleted.

Palatability

Good.

Limits to Usage (Anti-Nutritional Factors)

Many contain high levels of copper (50 ppm) as a result of the equipment used in the brewing industry. This may make barley distillers dark grains unsuitable for certain breeds of sheep. The oil being cereal derived is unsaturated.

Concentrate Inclusion % per species

	Inc %		Inc %		Inc %
Calf	20	Creep	0	Chick	0
Dairy	30	Weaner	0	Broiler	0
Beef	30	Grower	0	Breeder	0
Lamb	10	Finisher	0	Layer	0
Ewe	10	Sow	0		

Storage/Processing

Stores well.

Alternative Names

Barley Distillers, Malt Distillers.

Bulk Density

600Kg/m³

Typical Analysis

Dry Matter	90.0	NCDG	69.0	DUP (@ 5)	12.1	Avail Lysine	0.7
Crude Protein	26.0	NDF	50.3	DUP (@ 8)	13.5	Methionine	0.35
DCP	20.1	ADF	16.0	Salt	0.5	Meth & Cysteine	0.72
MER	12.7	Starch	2.6	Ca	0.15	Tryptophan	0.25
MEP	10.1	Sugar	3.7	Total Phos	0.90	Threonine	0.95
DE	11.1	Starch + Sugars	6.3	Av Phos	0.60	Arginine	1.1
Crude Fibre	13.6	FME	9.7	Magnesium	0.30	PDIA	10.3
Oil (EE)	9.1	ERDP (@ 2)	15.0	Potassium	0.94	PDIN	18.3
Oil (AH)	10.3	ERDP (@ 5)	12.1	Sodium	0.20	PDIE	15.6
EFA	3.5	ERDP (@ 8)	10.1	Chloride	0.32	Met DI	0.28
Ash	6.5	DUP (@ 2)	8.0	Total Lysine	0.9	Lys DI	0.11

Starch 2.6%	NDF 50.3%	Other 0.6%	
Sugars 3.7%	Ash 6.5%		
Protein 26%	Oil 10.3%		

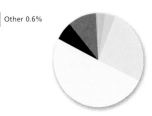

Cereals and By-Products

Introduction

By-product of alcohol distilling obtained by drying solid residues of fermented grain to which pot ale syrup or evaporated spent wash has been added.

A by-product of either grain whisky production (often for gin or vodka manufacture) or of ethanol production. Maize is soaked to release the starch reserves for fermentation. Often, some barley malt is added to start the process providing enzymes to convert starch to sugar. The grain which remains after the liquor is removed is often called wet draff. This can be pressed and dried with the left over yeast syrups to produce dark grains.

Origin/Place of Manufacture

USA, Brazil, UK.

Nutritional Benefit

High in fibre, but well digested by ruminants. A high-energy, mid-protein feed which is reasonably undegradable in the rumen. It is low in starch because of its extraction and has the highest energy value of all distillers grains, being higher in oil content than barley distillers dark grains.

Colour/Texture

UK produced - golden brown to dark brown. Imported - yellow/brown meal.

Palatability

Good.

Limits to Usage (Anti-Nutritional Factors)

The total level of distillers product in the diet should be considered for ruminants as they are low in starch and already pre-fermented. In common with all distillers by-products the residual cereal oil is unsaturated. This needs to be accounted for in ruminant formulations if other ingredients containing unsaturated oils are present.

Concentrate Inclusion % per species

	Inc %		Inc %		Inc %
Calf	0	**Creep**	0	**Chick**	0
Dairy	20	**Weaner**	2.5	**Broiler**	2.5
Beef	25	**Grower**	2.5	**Breeder**	5
Lamb	2.5	**Finisher**	5.0	**Layer**	5
Ewe	10.0	**Sow**	5		

Storage/Processing

Usually produced as a pellet although distillers is often a meal, either because it was not pelleted or has broken down during transit.

Alternative Names Corn distillers grains/meal.

Bulk Density Variable 350 - 600 kg/m³

Typical Analysis

Dry Matter	90.0	NCGD	82.0	DUP (@ 5)	11.5	Avail Lysine	0.5
Crude Protein	28.0	NDF	44.5	DUP (@ 8)	12.8	Methionine	0.50
DCP	20.0	ADF	14.5	Salt	0.2	Meth & Cysteine	0.85
MER	14.8	Starch	4.5	Ca	0.2	Tryptophan	0.2
MEP	12.9	Sugar	5.5	Total Phos	1.0	Threonine	1.1
DE	12.3	Starch + Sugars	10.0	Av Phos	0.5	Arginine	1.7
Crude Fibre	8.5	FME	11.0	Magnesium	0.4	PDIA	11.1
Oil (EE)	11.0	ERDP (@ 2)	16.7	Potassium	1.1	PDIN	20.1
Oil (AH)	12.0	ERDP (@ 5)	13.5	Sodium	0.15	PDIE	16.8
EFA	6.0	ERDP (@ 8)	12.2	Chloride	0.05	Met DI	0.32
Ash	4.9	DUP (@ 2)	9.0	Total Lysine	0.8	Lys DI	1.1

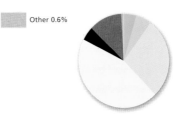

Starch 4.5%		NDF 44.5%		Other 0.6%	
Sugars 5.5%		Ash 4.9%			
Protein 28%		Oil 12%			

Cereals and By-Products	

Introduction

By-product of alcohol distilling obtained by drying solid residues of fermented grain to which pot ale syrup or evaporated spent wash has been added.

By-products of the grain distillery from whisky, gin and vodka production. Available with and without wheat solubles. Wheat is soaked to release the starch reserves for fermentation. Often, some barley malt is added to start the process, providing enzymes to convert starch to sugar. The grain which remains after the liquor is removed is often called wet draff. This can be pressed and dried with the left-over yeast syrups to produce dark grains.

Origin/Place of Manufacture

Mainly UK.

Nutritional Benefit

High in energy and protein which is partly degradable. Contains digestible fibre at moderate levels which makes it ideal for ruminants but less suitable for non-ruminants. Protein is partially rumen undegradable due to the heat used during the manufacturing process.

Colour/Texture

Dark brown meal/pellets.

Palatability

Good.

Limits to Usage (Anti-Nutritional Factors)

High copper levels (150 ppm) mean it is unsuitable for feeding to sheep. The free unsaturated oil content should be noted when used in ruminant diets.

Concentrate Inclusion % per species

	Inc %		Inc %		Inc %
Calf	10	Creep	0	Chick	0
Dairy	40	Weaner	0	Broiler	5
Beef	40	Grower	2.5	Breeder	5
Lamb	0	Finisher	5.0	Layer	5
Ewe	0	Sow	5.0		

Storage/Processing

Alternative Names

Wheat Distillers, Dried Grains.

Bulk Density

600 Kg/m³

Typical Analysis

Dry Matter	90.0	NCDG	80.0	DUP (@ 5)	25.0	Avail Lysine	0.8
Crude Protein	34.0	NDF	40.0	DUP (@ 8)	26.5	Methionine	0.55
DCP	26.0	ADF	18.0	Salt	0.55	Meth & Cysteine	0.9
MER	13.7	Starch	5.0	Ca	0.1	Tryptophan	0.3
MEP	11.1	Sugar	5.0	Total Phos	0.9	Threonine	1.2
DE	12.0	Starch + Sugars	10.0	Av Phos	0.6	Arginine	1.2
Crude Fibre	8.0	FME	11.0	Magnesium	0.35	PDIA	15.0
Oil (EE)	7.0	ERDP (@ 2)	26.7	Potassium	1.5	PDIN	25.5
Oil (AH)	8.5	ERDP (@ 5)	25.5	Sodium	0.1	PDIE	20.5
EFA	4.5	ERDP (@ 8)	24.9	Chloride	0.4	Met DI	0.4
Ash	5.0	DUP (@ 2)	16.5	Total Lysine	1.0	Lys DI	0.9

Starch 5%	NDF 40%	Other 2.5%
Sugars 5%	Ash 5%	
Protein 34%	Oil 8.5%	

Cereals and By-Products

Introduction

Maize or wheat is mashed to produce a liquor substrate which is fermented by yeast to produce alcohol spirit for further distillation. The remaining mixture is called draff and contains the insoluble and fibrous components of the grain, plus yeast and fermentation residues. It can be dried to produce dark grains, or sold moist as draff.

Origin/Place of Manufacture

As it contains high moisture levels, it is not economic to transport over long distances.

Nutritional Benefit

A mid-protein and mid-energy product. It is a succulent, palatable feed for ruminants. Lower in energy than cereals but higher in protein, which is of a less degradable nature. High levels of relatively undigestible fibre and oil (which adds to the energy value) are present. The oil is unsaturated which, at high levels, may reduce fibre digestion slightly. Ideal for sheep, dairy and beef rations, but unsuitable for young pigs and poultry.

Colour/Texture

Light brown and fibrous.

Palatability

Good when fresh and ideal for increasing combined forage intake.

Limits to Usage (Anti-Nutritional Factors)

High copper levels (60 mg/kg DM) from malt distillers mean care must be taken if considering feeding to sheep. Low in minerals except phosphorus.

Forage Inclusion % per species

	Inc %		Inc %		Inc %
Calf	5	Creep	0	Chick	0
Dairy	10	Weaner	0	Broiler	0
Beef	15	Grower	0	Breeder	0
Lamb	0	Finisher	5	Layer	0
Ewe	5	Sow	7.5		

Storage/Processing

Easily ensiled for winter feeding.

Alternative Names

Bulk Density

Typical Analysis

Dry Matter	23.0	NCGD	60.0	DUP (@ 5)	5.9	Avail Lysine	-
Crude Protein	25.0	NDF	57.5	DUP (@ 8)	7.0	Methionine	0.45
DCP	20.0	ADF	32.0	Salt	0.5	Meth & Cysteine	0.95
MER	11.6	Starch	3.2	Ca	0.2	Tryptophan	0.25
MEP	-	Sugar	1.8	Total Phos	0.51	Threonine	0.7
DE	-	Starch + Sugars	5.0	Avail Phos	-	Arginine	1.0
Crude Fibre	20.0	FME	9.0	Magnesium	0.17	PDIA	13.9
Oil (EE)	6.0	ERDP (@ 2)	16.0	Potassium	0.06	PDIN	18.9
Oil (AH)	6.5	ERDP (@ 5)	13.5	Sodium	0.03	PDIE	18.0
EFA	-	ERDP (@ 8)	12.5	Chloride	0.05	Met DI	0.35
Ash	4.0	DUP (@ 2)	3.5	Total Lysine	0.8	Lys DI	0.6

Starch 3.2%	NDF 57.5%	Other 2%
Sugars 1.8%	Ash 4%	
Protein 25%	Oil 6.5%	

Miscellaneous

Introduction

Usually a calcium or magnesium soap manufactured from palm oil (palm fatty acid distillate), although tallow has been used in America.

Oils can coat the fibre and reduce its availability to the rumen microbes further, some free fatty acids can be toxic to certain rumen bacteria. Either action, or both combined could seriously disrupt rumen function. Hard fats have less effect than soft fats, with the formation of a soap making the material more inert. A palm fatty acid product with either calcium or magnesium, is used to give the fat 'protection', allowing it to pass to the abomasum and into the small intestine for digestion.

Origin/Place of Manufacture

UK, Europe, USA.

Nutritional Benefit

Provides the most concentrated form of energy available to ruminants and is ideal to make up energy deficit in early lactation. Especially useful for large cows in early lactation or a high yielding animal where energy density is required.

Colour/Texture

Cream/Grey.

Palatability

No problem when mixed with other feeds.

Limits to Usage (Anti-Nutritional Factors)

High levels may raise butter fats in ruminant animals.

Concentrate Inclusion % per species

	Inc %		Inc %		Inc %
Calf	0	Creep	0	Chick	0
Dairy	5	Weaner	0	Broiler	0
Beef	3	Grower	0	Breeder	0
Lamb	0	Finisher	0	Layer	0
Ewe	5	Sow	0		

Storage/Processing

Stores well in dry store.

Alternative Names

Various commercial trade names exist.

Bulk Density

Typical Analysis

Dry Matter	99.0	NCGD	63.0	DUP (@ 5)	0	Avail Lysine	0
Crude Protein	0	NDF	0	DUP (@ 8)	0	Methionine	0
DCP	0	ADF	0	Salt	0	Meth & Cysteine	0
MER	31.0	Starch	0	Ca	9.5	Tryptophan	0
MEP	35.0	Sugar	0	Total Phos	0	Threonine	0
DE	37.0	Starch + Sugars	0	Av Phos	0	Arginine	0
Crude Fibre	0	FME	0	Magnesium	0	PDIA	0
Oil (EE)	86.0	ERDP (@ 2)	0	Potassium	0	PDIN	0
Oil (AH)	86.0	ERDP (@ 5)	0	Sodium	0	PDIE	9.2
EFA	12.0	ERDP (@ 8)	0	Chlorine	0	Met DI	0
Ash	14.0	DUP (@ 2)	0	Total Lysine	0	Lys DI	0

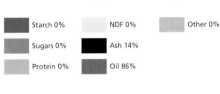

Starch 0%   NDF 0%   Other 0%

Sugars 0%   Ash 14%

Protein 0%   Oil 86%

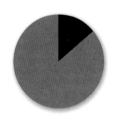

Miscellaneous

Introduction

These are fed in the ration to provide energy and are derived from a variety of sources, which include soya bean, palm, tallow, fish oils and their derivatives.
Care should be taken to ensure only high quality oils are used.

Origin/Place of Manufacture

UK, Europe, USA.

Nutritional Benefit

Fats have three times the energy value of cereals and are therefore an ideal way of meeting energy demand. They also supply essential fatty acids. Specialist products are available to be rumen inert eg. fat prills have a bypass effect for ruminants due to the low melting point of the fats used.

Colour/Texture

Golden to grey.

Palatability

Poor.

Limits to Usage

Soft oils will reduce digestion and could produce soft fats in the animal. Harder oils, eg. palm derivatives, are more difficult for non-ruminants to digest. Black products should be avoided.

Concentrate Inclusion % per species

	Inc %		Inc %		Inc %
Calf	2.5	Creep	5	Chick	5
Dairy	3	Weaner	4	Broiler	5
Beef	3	Grower	4	Breeder	5
Lamb	2	Finisher	2	Layer	2.5
Ewe	3	Sow	2.5		

Storage/Processing

Depending on the product's melting point, the storage tank may need to be heated.

Alternative Names

Bulk Density

Typical Analysis	Fat 50%	Fat 100% GP	Fat Flakes	Typical Analysis	Fat 50%	Fat 100% GP	Fat Flakes
Dry Matter	95.0	99.0	99.0	DUP (@ 5)	1.1	0	0
Crude Protein	5.0	0	0	DUP (@ 8)	1.3	0	0
DCP	4.0	0	0	Salt	0.1	0	0
MER	22.0	38.0	36.0	Ca	0.2	0	0
MEP	21.0	35.0	36.0	Total Phos	0.3	0	0
DE	29.5	39.0	40.0	Av Phos	0.1	0	0
Crude Fibre	8.5	0	0	Magnesium	0.2	0	0
Oil (EE)	55.0	98.5	99.0	Potassium	0.5	0	0
Oil (AH)	55.0	99.0	99.0	Sodium	0.02	0	0
EFA	11.0	20.0	20.0	Chlorine	0.1	0	0
Ash	8.0	1.0	1.0	Total Lysine	0.35	0	0
NCGD	80.0	95.0	95.0	Avail Lysine	0.20	0	0
NDF	0	0	0	Methionine	0.15	0	0
ADF	0	0	0	Meth & Cysteine	0.25	0	0
Starch	2.0	0	0	Tryptophan	0.1	0	0
Sugar	2.0	0	0	Threonine	0.25	0	0
Starch + Sugars	4.0	0	0	Arginine	0.8	0	0
FME	3.2	0	0	PDIA	0	0	0
ERDP (@ 2)	4.6	0	0	PDIN	0	0	0
ERDP (@ 5)	4.1	0	0	PDIE	4.7	9.2	9.2
ERDP (@ 8)	3.6	0	0	Met DI	0	0	0
DUP (@ 2)	0.7	0	0	Lys DI	0	0	0

- Starch 0%
- Sugars 0%
- Protein 0%
- NDF 0%
- Ash 1%
- Oil 99%
- Other 0%

Miscellaneous	

Introduction

Products obtained by hydrolysing, drying and grinding poultry feathers.

Feathers from the poultry industry are indigestible and they contain keratins. However, when cooked under extreme conditions of temperature and pressure, dried and ground, the protein becomes partly available.

Origin/Place of Manufacture

UK and around the world.

Nutritional Benefit

High in protein but of low quality. It is missing many of the essential amino acids, and only half of the protein is available. Large differences occur between turkey, old hens and broilers, due to the varying levels of skin, blood, etc., present. More suited to ruminants than to monogastrics. Especially low in lysine and methionine.

Colour/Texture

Light brown friable meal.

Palatability

Unpalatable.

Limits to Usage (Anti-Nutritional Factors)

Needs balancing with Lysine and Methionine.

Concentrate Inclusion % per species

	Inc %		Inc %		Inc %
Calf	0	Creep	0	Chick	0
Dairy	2.5	Weaner	0	Broiler	0
Beef	7.5	Grower	0	Breeder	0
Lamb	3	Finisher	0	Layer	0
Ewe	2.5	Sow	0		

Storage/Processing

Alternative Names

Hydrolysed Feather Meal.

Bulk Density

475 - 550 kg/m³

Typical Analysis

Dry Matter	90.0	NCGD	83.0	DUP (@ 5)	41.8	Avail Lysine	1.8
Crude Protein	86.0	NDF	0	DUP (@ 8)	45.3	Methionine	0.8
DCP	68.0	ADF	0	Salt	0.4	Met & Cysteine	4.9
MER	14.1	Starch	0.2	Ca	0.60	Tryptophan	0.59
MEP	14.2	Sugar	0.5	Total Phos	0.75	Threonine	4.2
DE	14.2	Starch + Sugars	0.7	Av Phos	0.69	Arginine	6.0
Crude Fibre	0.6	FME	11.8	Magnesium	0.25	PDIA	40.7
Oil (EE)	6.5	ERDP (@ 2)	32.3	Potassium	0.25	PDIN	65.1
Oil (AH)	6.5	ERDP (@ 5)	23.0	Sodium	0.2	PDIE	43.0
EFA	0.3	ERDP (@ 8)	20.0	Chloride	0.2	Met DI	0.3
Ash	4.0	DUP (@ 2)	30.8	Total Lysine	2.5	Lys DI	1.2

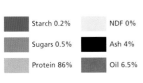

Starch 0.2% NDF 0% Other 2.8%

Sugars 0.5% Ash 4%

Protein 86% Oil 6.5%

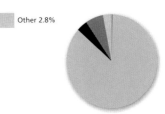

Miscellaneous	

Introduction

Product obtained by processing whole or parts of fish from which part of the oil may have been removed and to which fish solubles may have been re-added.

Either produced as a by-product of fish processing for human consumption or as a direct product for animal feed when demand is high. Around 6.5 million tonnes of fishmeal and 1.2 million tonnes of fish oil are produced annually worldwide from around 30 million tonnes of fish. Fishmeal is produced from fish/fish trimmings which are cooked/separated from the oil, dried, pressed, ground and treated, to form a meal. Whole fish from carefully managed stocks of species deemed unsuitable for human consumption are the main constituent of fishmeals.

Origin/Place of Manufacture

Manufactured in the UK, Ireland, S. America and Europe. Large quantities from S. America, especially Peru, which is the largest single producer of fishmeal in the world, Chile being the second. The three main types are White (UK), Chilean and Herring.

(Northern European production). **British** - Good quality. **Peruvian** - Generally Anchovy and Sardines. **Chilean** - Generally Anchovy and Jack Mackeral. **Icelandic** - Generally Capelin and Scania Herring. **Scandinavian** - Generally Sand Eel, Blue Whiting and Sprat.

Nutritional Benefit

An excellent protein source. Quality will vary, depending on manufacturing process, the time period between catching and manufacture, or the freshness of the raw material. Fresh Raw Material implies a lower degree of amino acid breakdown leading to better palatability and digestibility of the fishmeal.

Generally, the more gentle the drying process, the higher the quality of the meal. Direct flame drying, which is regarded as the fastest form of drying, is also the most damaging. Higher temperatures usually lead to a higher percentage of the protein fraction in the meal being damaged. The best method of drying is believed to be indirect hot air drying which, although slower than flame drying, generally produces a better quality final product.

Flame dried materially usually has a higher ash level and would generally not be the first choice for a poultry producer. Fishmeal has a good amino acid profile for all animals and is high in Undegradable Protein for ruminants. Protein may vary from 60-70% between products. Rich in lysine, sulphur, minerals and vitamins. Blended products may have lower nutritional quality than straight, single source, fishmeal.

Colour/Texture

Grey to brown meal. (More direct heat results in a darker colour).

Palatability

Average.

Limits to Usage (Anti-Nutritional Factors)

Milk taint may result when the dairy ration contains high levels of fish oil. Contains high levels of ash. High levels in poultry may cause gizzard crown and egg taint.

Concentrate Inclusion % per species

	Inc %		Inc %		Inc %
Calf	5	Creep	10	Chick	5
Dairy	5	Weaner	10	Broiler	5
Beef	5	Grower	7.5	Breeder	5
Lamb	5	Finisher	4	Layer	2.5
Ewe	5	Sow	3		

Storage/Processing

Stores well.

Alternative Names

Capelin Meal, Herring Meal.

Bulk Density

550 - 650 kg/m³

Typical Analysis	Chilean	Herring	Whiter	Scottish
Dry Matter	91.0	91.0	91.0	91.0
Crude Protein	73.0	77.0	72.0	73.0
DCP	69.0	72.0	69.0	69.0
MER	14.5	14.8	11.1	14.6
MEP	17.0	17.9	16.3	16.3
DE	17.0	15.8	16.0	16.2
Crude Fibre	0.5	0.5	0.5	0.5
Oil (EE)	10.0	10.0	6.0	11.0
Oil (AH)	11.0	11.0	7.0	11.0
EFA	4.5	5.0	3.0	4.0
NCGD	87.0	87.0	79.0	82.0
NDF	3.5	3.5	3.5	3.5
ADF	0.8	0.8	0.8	0.8
Starch	0	0	0	0
Sugar	0.1	0.1	0.1	0.1
Starch + Sugars	0.1	0.1	0.1	0.1
FME	9.8	11.8	10.0	10.2
ERDP (@ 2)	44.5	48.0	40.5	41.0
ERDP (@ 5)	34.5	37.2	30.5	31.0
ERDP (@ 8)	29.5	29.0	26.5	27.0
DUP (@ 2)	9.8	16.2	16.9	12.2
DUP (@ 5)	19.5	28.0	26.0	26.4
DUP (@ 8)	24.0	32.0	29.5	29.8
Salt	2.5	2.2	2.5	2.5
Ca	3.5	2.7	6.8	3.0
Total Phos	2.7	2.7	3.5	3.5
Av Phos	2.3	2.1	3.3	3.0
Magnesium	0.3	0.2	0.3	0.3
Potassium	1.3	1.3	1.0	1.1
Sodium	1.0	0.8	1.1	1.1
Chloride	1.5	1.5	1.55	1.6
Total Lysine	5.5	6.1	4.8	4.8
Avail Lysine	5.3	5.9	4.6	4.6
Methionine	1.8	2.1	1.9	1.9
Meth & Cysteine	2.7	2.9	2.6	2.6
Tryptophan	0.9	0.9	0.7	0.7
Threonine	2.9	3.4	2.8	2.5
Arginine	3.8	4.1	4.4	4.3
PDIA	37.2	40.1	36.7	36.6
PDIN	53.1	57.5	54.2	54.0
PDIE	39.1	42.5	37.3	37.0
Met DI	1.0	1.1	0.9	0.9
Lys DI	3.1	3.2	3.2	3.2

Starch 0%   NDF 3.5%   Other 0%

Sugars 0.1%   Ash 12.4%

Protein 73%   Oil 11%

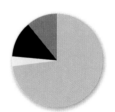

Forages and Stock Feeds

Introduction

Fodder beet has a potentially higher yield of digestible nutrients than any other forage crop. The best crops may give 18 tonnes of dry matter per hectare. However, it is a crop which demands 'arable expertise' and there can be problems with late harvesting, cleaning and the feeding process. Specialist equipment for cleaning and chopping is now available and may help to increase the interest in fodder beet on many farms. Sown May/June for harvesting Oct. Dec.

Origin

UK, Ireland, Denmark, Netherlands.

Nutritional Benefit

Can produce more dry matter/acre than cereal grains. A sugar rich energy feed for ruminants but the composition can vary. Soil contamination must be avoided to prevent digestive upsets as should excessive feeding; in severe cases, this can cause hypocalcaemia and even death. The digestive upsets are due to excess sugar in the rumen, and/or mineral imbalance. Chopping enhances intake in ruminants and cattle relish the root due to its succulence and sugar content.

Colour/Texture

Grey white fleshy tuber.

Palatability

Excellent.

Limits to Usage (Anti-Nutritional Factors)

Fodder beet should be well cleaned, preferably at harvesting but certainly before feeding. Fodder beet tops can also be fed and are of a lower dry matter (12%) and energy (10 MJ/kg DM) but higher protein (16%). Tops should be wilted to avoid metabolic and digestive upsets. Root tops can be associated with milk taint in dairy cows. High FME may limit inclusion.

Forage Inclusion % per species

Ewes (70 Kg liveweight) - 2.5 kg/head/day.
Beef should be limited to 3.5 kg/100 kg liveweight.
Cows - early lactation 1.7 kg/100kg liveweight.
- mid/late lactation 3.0 kg/100 kg liveweight.

	Inc %		Inc %		Inc %
Calf	10	Creep	0	Chick	0
Dairy	20	Weaner	0	Broiler	0
Beef	20	Grower	0	Breeder	0
Lamb	15	Finisher	0	Layer	0
Ewe	20	Sow	0		

Storage/Processing Store on a dry, concrete apron if possible.

Alternative Names

Bulk Density

Typical Analysis

Dry Matter	18.0	NCGD	87.0	DUP (@ 5)	0.7	Avail Lysine	-
Crude Protein	7.0	NDF	19.5	DUP (@ 8)	1.0	Methionine	0.7
DCP	6.0	ADF	9.5	Salt	0.7	Meth & Cysteine	-
MER	12.5	Starch	2.0	Ca	0.3	Tryptophan	-
MEP	-	Sugar	64.7	Total Phos	0.25	Threonine	-
DE	-	Starch + Sugars	66.7	Avail Phos	0.2	Arginine	-
Crude Fibre	6.0	FME	11.7	Magnesium	0.15	PDIA	0.7
Oil (EE)	0.7	ERDP (@ 2)	5.2	Potassium	1.5	PDIN	3.6
Oil (AH)	0.8	ERDP (@ 5)	4.9	Sodium	0.3	PDIE	8.3
EFA	-	ERDP (@ 8)	4.7	Chloride	0.4	Met DI	0.6
Ash	6.0	DUP (@ 2)	0.5	Total Lysine	0.3	Lys DI	0.2

- Starch 2%
- NDF 19.5%
- Other 0%
- Sugars 64.7%
- Ash 6%
- Protein 7%
- Oil 0.8%

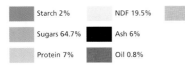

Roots, Fruits and By-Products

Introduction

The by-product of the processing of grapes Vitis vinifera L. after the juice has been pressed out.

The pulp remaining after grape juice extraction. Normally, pulp consists of 60% pulp and 40% seeds.

Origin/Place of Manufacture

Grown around the world, especially in Europe: Italy, Spain and Germany.

Nutritional Benefit

Similar in digestible energy to nutritionally improved straw (NIS) and useful as a feed extender. Contains digestible cell walls, making it high in fibre, low in protein and energy.

Colour/Texture

Green/brown, meal/pellet.

Palatability

Average.

Limits to Usage (Anti-Nutritional Factors)

Tannins and copper levels may be high.

Concentrate Inclusion % per species

	Inc %		Inc %		Inc %
Calf	0	Creep	0	Chick	0
Dairy	10	Weaner	0	Broiler	0
Beef	10	Grower	0	Breeder	0
Lamb	0	Finisher	0	Layer	0
Ewe	5	Sow	0		

Storage/Processing

Alternative Names

Bulk Density

Typical Analysis

Dry Matter	86.0	NCDG	-	DUP (@ 5)	-	Avail Lysine	-
Crude Protein	12.0	NDF	68.0	DUP (@ 8)	-	Methionine	-
DCP	-	ADF	-	Salt	0.2	Meth & Cysteine	-
MER	5.5	Starch	0	Ca	0.8	Tryptophan	-
MEP	-	Sugar	0	Total Phos	0.15	Threonine	-
DE	-	Starch + Sugars	0	Av Phos	0.05	Arginine	-
Crude Fibre	26.0	FME	-	Magnesium	1.0	PDIA	-
Oil (EE)	4.0	ERDP (@ 2)	-	Potassium	-	PDIN	-
Oil (AH)	4.1	ERDP (@ 5)	-	Sodium	0.05	PDIE	-
EFA	-	ERDP (@ 8)	-	Chlorine	0.1	Met DI	-
Ash	10.0	DUP (@ 2)	-	Total Lysine	-	Lys DI	-

Starch 0% NDF 68% Other 5.9%

Sugars 0% Ash 10%

Protein 12% Oil 4.1%

Forages and Stock Feeds

Introduction

Grassland can be divided into two main groups: hill and rough grassland and cultivated grassland (permanent and temporary pastures). Fresh grass remains the main feed for ruminants during the spring, summer and autumn, with conserved grass (silage, hay) being fed during the winter.

Origin/Place of Manufacture

Worldwide.

Nutritional Benefit

The nutritional value of grass is variable depending on species in sward, location, weather, time of year, ley, age and fertiliser application rates/timings. The crude protein can range from 4% - 30% in heavily fertilised pastures, with crude fibre content ranging from 20% - 45% in very mature samples.

Early in the growing season, it has a high water, organic acid and protein content with low content of carbohydrates and lignin making it highly digestible. As the plant matures, the yield of forage increases; there is increased structural carbohydrate and lignin, and decreased protein and energy levels.

The moisture content can vary dramatically, being highest in the early stages of growth (80-85%) and lowest when the plant seeds. The weather also greatly affects moisture level.

Soluble carbohydrates range from 4 -30% with the highest levels found in varieties such as Italian ryegrass. The cellulose and hemicellulose content varies between 20-30% of DM and 10-30% of DM respectively. Both cellulose and hemicellulose increases as the plant matures, as does lignin, which influences rumen digestibility.

The protein content decreases with maturity and the amino acid profile does not alter much with arginine, glutamic acid and lysine present in reasonable levels. The non-protein nitrogen (NPN) content and nitrogen level both decrease as the plants mature, with levels tending to be higher during good growing conditions.

Oil levels are relatively low in grass and usually below 4% (of DM) made up largely of unsaturated fatty acids.

Mineral/vitamin content varies depending on soil type, stage of growth and fertiliser application. Grass is rich in carotene, a precursor of vitamin A, which is present in large amounts.

Palatability Good.

Limits to Usage (Anti-Nutritional Factors)

Stocking rates, sward height and density, parlour feed substitution rates, trace element imbalances and high molybdenum contents in improved or 'teart' pastures.
Later in the season grazing may provide insufficient energy for high performing animals, particularly dairy cows, and careful supplementation is required.

Forage Inclusion % per species

Depends largely on production level required. Minimum inclusion rate for ruminants if sole forage is approx 40%.

	Inc %		Inc %		Inc %
Calf	100	Creep	0	Chick	0
Dairy	100	Weaner	0	Broiler	0
Beef	100	Grower	0	Breeder	0
Lamb	100	Finisher	0	Layer	0
Ewe	100	Sow	0		

Storage/Processing

Can be harvested and fed in zero grazed system, maximising output per acre.
Grass is usually conserved as silage and/or hay.

Bulk Density

Typical Analysis

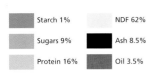

Dry Matter	18.0	NDF	62.0	DUP (@ 8)	4.7	Methionine	0.3
Crude Protein	16.0	ADF	30.0	Salt	0.2	Meth & Cysteine	0.5
DCP	11.0	Starch	1.0	Ca	0.7	Tryptophan	0.2
MER	11.3	Sugar	9.0	Total Phos	0.4	Threonine	0.7
MEP	4.8	Starch + Sugars	10.0	Av Phos	0.3	Arginine	0.7
DE	8.2	FME	7.9	Magnesium	0.2	PDIA	4.9
Crude Fibre	20.0	ERDP (@ 2)	11.6	Potassium	0.5	PDIN	10.0
Oil (EE)	3.3	ERDP (@ 5)	9.4	Sodium	0.1	PDIE	9.2
Oil (AH)	3.5	ERDP (@ 8)	8.3	Chloride	0.1	Met DI	0.2
Ash	8.5	DUP (@ 2)	1.8	Total Lysine	0.7	Lys DI	0.65
NCGD	96.0	DUP (@ 5)	3.7	Av Lysine	-		

Starch 1% NDF 62% Other 0%

Sugars 9% Ash 8.5%

Protein 16% Oil 3.5%

Grass Hay

Forages and Stock Feeds

Introduction

Historically, hay was the most common method of conserving grass. It aims to reduce the moisture content of the green crop to a level low enough to inhibit the action of plant and microbial enzymes. Hay can be made from traditional pastures or meadows, or a residual hay from a grass seed crop.

Origin/Place of Manufacture

Around the world.

Nutritional Benefit

The quality of conserved grass is obviously dependent on the quality of the grass used. For adequate conservation, the moisture content of grass must be reduced to approximately 15%. The plants inevitably lose some nutrients from the action of plant and microbial enzymes, chemical oxidation, leaching and mechanical damage. Quicker drying, therefore, reduces the loss of nutrients from the fresh forage. Nutritional quality of the hay is obviously dependent on the quality of the fresh forage (see Grass Fresh), and the losses listed above plus any residual respiration losses in storage.

Colour/Texture

Green and fibrous.

Palatability

Good, if well preserved.

Limits to Usage (Anti-Nutritional Factors)

The nutritive value varies dramatically, so appropriate analysis should be undertaken. Badly made or stored crops can be dusty and or mouldy. Check for poisonous weeds if origin of bales is unknown.

Forage Inclusion % by species

	Inc %		Inc %		Inc %
Calf	100	Creep	0	Chick	0
Dairy	100	Weaner	0	Broiler	0
Beef	100	Grower	0	Breeder	0
Lamb	100	Finisher	0	Layer	0
Ewe	100	Sow	0		

Storage/Processing

Drying can be speeded up by conditioners. However, excessive mechanical damage can promote further nutrient loss.

Bulk Density

Typical Analysis

Dry Matter	87.0	NCGD	57.0	DUP (@ 5)	3.5	Avail Lysine	-
Crude Protein	10.4	NDF	68.8	DUP (@ 8)	5.0	Methionine	0.15
DCP	6.0	ADF	38.0	Salt	0.1	Meth & Cysteine	0.3
MER	8.5	Starch	0.3	Ca	0.25	Tryptophan	0.15
MEP	-	Sugar	11.0	Total Phos	0.25	Threonine	0.4
DE	-	Starch + Sugars	11.3	Av Phos	-	Arginine	0.4
Crude Fibre	33.0	FME	8.1	Magnesium	0.2	PDIA	2.8
Oil (EE)	1.5	ERDP (@ 2)	7.0	Potassium	2.0	PDIN	6.7
Oil (AH)	2.0	ERDP (@ 5)	5.6	Sodium	0.05	PDIE	7.7
EFA	-	ERDP (@ 8)	5.0	Chloride	0.05	Met DI	0.1
Ash	7.5	DUP (@ 2)	2.0	Total Lysine	0.3	Lys DI	0.2

- Starch 0.3%
- Sugars 11%
- Protein 10.4%
- NDF 68.8%
- Ash 7.5%
- Oil 2%
- Other 0%

CONTEXT

Forages and Stock Feeds	

Introduction

Product obtained by drying and milling young forage plants.

Manufactured by drying the fresh forage at high temperatures (800°C) for a short period (60 seconds) in a large volume of air, to reduce the moisture from 80% to 10%, whilst maintaining the feed value. Usually manufactured from 'Italian' and perenial ryegrass with lucerne (alfalfa) increasing in importance.
(Clover and sanfoin are also used). It is usually rough ground and then pelleted.

Origin/Place of Manufacture

UK, Europe.

Nutritional Benefit

Rich in crude protein (17% home produced, 15% continental) and digestible fibre.
A good feed which retains most of the nutritional value of grass with some claims of enhancement. Quality depends on grass and the stage at which it was cut. Does not depress rumen pH, so ideal with cereals. Contains beta- carotene. The drying process has been claimed to reduce the degradability of the protein, providing more nutrients for digestion in the small intestine (DUP). 85-90% of the Metabolisable Energy is also available for fermentation in the rumen.

UK production tends to be higher in protein levels than imported meal/pellets which may have been sun-dried and of lower nutritional quality.

Colour/Texture

Green meal/pellet.

Palatability

Good.

Limits to Usage (Anti-Nutritional Factors)

If finely ground, it can reduce milk fat content of milk. Avoid dark products where over-heating may affect the protein quality.

Concentrate Inclusion % per species

	Inc %		Inc %		Inc %
Calf	10	Creep	0	Chick	0
Dairy	30	Weaner	0	Broiler	0
Beef	30	Grower	2.5	Breeder	5
Lamb	15	Finisher	2.5	Layer	5
Ewe	30	Sow	2.5		

Storage/Processing

Alternative Names

Bulk Density

Nuts 600 - 650 kg/m³
Meal 250 - 275 kg/m³

Typical Analysis

Dry Matter	90.0	NCGD	92.0	DUP (@ 5)	3.9	Avail Lysine	0.7
Crude Protein	17.0	NDF	54.5	DUP (@ 8)	5.3	Methionine	0.3
DCP	11.6	ADF	28.0	Salt	0.3	Meth & Cysteine	0.45
MER	10.8	Starch	1.5	Ca	1.0	Tryptophan	0.3
MEP	6.0	Sugar	12	Total Phos	0.5	Threonine	0.8
DE	8.5	Starch + Sugars	13.5	Av Phos	0.4	Arginine	0.8
Crude Fibre	24.0	FME	8.0	Magnesium	0.3	PDIA	5.5
Oil (EE)	3.5	ERDP (@ 2)	12.2	Potassium	2.6	PDIN	11.2
Oil (AH)	4.0	ERDP (@ 5)	9.9	Sodium	0.3	PDIE	9.5
EFA	1.0	ERDP (@ 8)	9.1	Chloride	0.1	Met DI	0.2
Ash	11.0	DUP (@ 2)	1.8	Total Lysine	0.9	Lys DI	0.7

Starch 1.5%	NDF 54.5%	Other 0%	
Sugars 12%	Ash 11%		
Protein 17%	Oil 4%		

CONTEXT

Grass Silage

Forages and Stock Feeds

Introduction

This remains the main form of conserved forage in the UK and Ireland. Material is harvested, then clamped and sealed to remove air. This encourages an aerobic fermentation, producing acids (primarily lactic) which reduce the pH to 3.8-4.4. Sometimes additives are used to assist the fermentation.

Origin/Manufacture

UK, Ireland.

Nutritional Benefit

Quality varies dramatically and material should always be analysed.
The dry matter is normally in the range 19-35% and crude protein 10-20%. Metabolisable energy of the grass silage is normally between 8.0 and 12.5 MJ/Kg DM. Very high metabolisable energy silages often do not perform as well as the analyses show, due to high rates of passage.

pH is normally in the range 3.8-4.2 with higher values in high dry matter restricted fermentation crops ie Big bale silage. A high pH (4.5-5.0) with low dry matter crops indicates a poor butyric or secondary fermentation. A high pH with a high dry matter crop merely indicates a restricted fermentation.

High levels of ammonia nitrogen, greater than 15% of total N are usually associated with poor fermentation. The ash level is normally below 10%. Values above this indicate potential soil contamination and poor fermentation. Silage normally has a low water soluble carbohydrate level due to fermentation, and is normally in the range of 3-10%. Residual sugars in wilted crops will be higher due to the restricted fermentation. Silage D value depends on growth stage at cutting and losses during the ensilage process. After the first cut, the aim is to cut six weekly regrowths for D65-D70. There is a tendency for D value to reduce in later season due to composition changes in the grass.

Colour/Texture

Green; leafy to fibrous depending on species and growth stage at cutting.

Palatability

Varies dramatically, but if well preserved, palatability is good.

Limits to Usage

The nutritive value varies dramatically so appropriate analysis should be undertaken. Wet silages limit the dry matter intakes (DMIs).

Forage Inclusion % per Species

Silage usually makes up between 40% and 100% of the dry matter intake of Dairy, Beef and Sheep.

	Inc %		Inc %		Inc %
Calf	100	Creep	0	Chick	0
Dairy	100	Weaner	0	Broiler	0
Ewe	100	Grower	0	Breeder	0
Lamb	100	Finisher	0	Layer	0
Beef	100	Sow	0		

Storage/Processing

Anaerobic condition must be maintained. This is achieved by chopping the crop during harvesting, rapid silo filling, adequate consolidation and sealing. A biological or acid additive can be added to improve fermentation.

Alternative Names

Bulk Density

600 - 700 kg/m³

Typical Analysis

Dry Matter	24.0	NDF	54.0	DUP (@ 8)	1.5	Methionine	0.2
Crude Protein	13.7	ADF	36.0	Salt	0.5	Meth & Cysteine	0.3
DCP	9.3	Starch	1.0	Ca	0.5	Tryptophan	0.2
MER	11.0	Sugar	2.0	Total Phos	0.3	Threonine	0.6
MEP	-	Starch + Sugars	3.0	Avail Phos	-	Arginine	0.5
DE	-	FME	8.0	Magnesium	0.2	PDIA	2.0
Crude Fibre	28.0	ERDP (@ 2)	10.0	Potassium	2.5	PDIN	8.0
Oil (EE)	3.0	ERDP (@ 5)	9.5	Sodium	0.2	PDIE	7.0
Oil (AH)	3.5	ERDP (@ 8)	9.0	Chloride	0.4	Met DI	0.15
Ash	8.0	DUP (@ 2)	0.5	Total Lysine	0.5	Lys DI	0.2
NCGD	69.0	DUP (@ 5)	1.0	Avail Lysine	-		

Big Bale

Typical Analysis

Dry Matter	35.0	NDF	50.0	DUP (@ 8)	3.3	Methionine	0.2
Crude Protein	12.5	ADF	38.0	Salt	2.0	Meth & Cysteine	0.3
DCP	10.0	Starch	1.0	Ca	0.5	Tryptophan	0.2
MER	10.5	Sugar	7.0	Total Phos	0.3	Threonine	0.6
MEP	-	Starch + Sugars	8.0	Avail Phos	-	Arginine	0.5
DE	-	FME	8.0	Magnesium	0.15	PDIA	2.1
Crude Fibre	29.0	ERDP (@ 2)	7.5	Potassium	2.5	PDIN	7.0
Oil (EE)	2.0	ERDP (@ 5)	7.4	Sodium	0.2	PDIE	6.9
Oil (AH)	2.5	ERDP (@ 8)	7.2	Chloride	0.4	Met DI	0.15
Ash	9.0	DUP (@ 2)	2.7	Total Lysine	0.5	Lys DI	0.2
NCGD	67.0	DUP (@ 5)	3.0	Avail Lysine	-		

Starch 1%　　NDF 54%　　Other 17.8%

Sugars 2%　　Ash 8%

Protein 13.7%　　Oil 3.5%

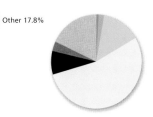

Oilseeds and By-Products	

Introduction

By-product of oil manufacture, obtained by expelling and/or extraction of partially decorticated groundnuts (Maximum crude fibre content 16% in the dry matter).

Groundnuts, often known as peanuts or monkey nuts, are dehulled and crushed for their oil for human consumption. The seed contains 25-30% Crude Protein and 35-60% oil. When crushed, the meal contains approximately 7% oil. This is reduced further if an extraction process is used.

Origin/Place of Manufacture

Grown in sub-tropical countries, eg. China, India and even USA.

Nutritional Benefit

A good source of protein and energy. Less degradable than rape meal, but more so than soya bean meal. Higher in fibre and energy and lower in protein and of a poorer quality (lower in lysine and methionine) than soya bean meal, with the analysis varying widely by source. Palatability can be reduced and the meal can even be toxic as it oxidises in warm, humid conditions at origin. The major concern in the past has been due to fungal attack of seeds and meal and resultant toxins produced. Deficient in Vitamin B12.

Colour/Texture

Mid brown pellets, cake or meal.

Palatability

Average.

Limits to Usage (Anti-Nutritional Factors)

Aflatoxin contamination is common. Specialist plants can potentially detoxify the material. Undecorticated meals have a lower energy value.

Concentrate Inclusion % per species

	Inc %		Inc %		Inc %
Calf	5	Creep	0	Chick	0
Dairy	15	Weaner	2.5	Broiler	2.5
Beef	15	Grower	2.5	Breeder	4.0
Lamb	5	Finisher	2.5	Layer	4.0
Ewe	10	Sow	2.5		

Storage/Processing

Needs careful storage and shipping to avoid aflatoxin contamination.

Alternative Names

Peanut/Monkey Nut Meal/Cake.

Bulk Density

Meal 525 - 700 kg/m³

Typical Analysis

Dry Matter	87.0	NCGD	80.1	DUP (@ 5)	8.0	Avail Lysine	1.4
Crude Protein	52.0	NDF	19.5	DUP (@ 8)	10.0	Methionine	0.6
DCP	45.0	ADF	14.0	Salt	0.1	Meth & Cysteine	1.2
MER	12.9	Starch	7.0	Ca	0.2	Tryptophan	0.5
MEP	12.9	Sugar	7.0	Total Phos	0.7	Threonine	1.4
DE	17.8	Starch + Sugars	14.0	Av Phos	0.3	Arginine	5.2
Crude Fibre	9.0	FME	11.5	Magnesium	0.5	PDIA	23.0
Oil (EE)	6.5	ERDP (@ 2)	43.2	Potassium	1.5	PDIN	36.0
Oil (AH)	7.5	ERDP (@ 5)	35.1	Sodium	0.03	PDIE	25.0
EFA	0.5	ERDP (@ 8)	33.0	Chloride	0.02	Met DI	0.3
Ash	7.0	DUP (@ 2)	3.4	Total Lysine	1.6	Lys DI	1.2

- Starch 7%
- NDF 19.5%
- Other 0%
- Sugars 7%
- Ash 7%
- Protein 52%
- Oil 7.5%

Guar Meal

44

Legumes and By-Products	

Introduction

By-product obtained after extraction of the mucilage from seeds of Cyamopsis tetragonoloba (L.) Taub.

Produced from the Guar, a legume grown for the gum found in the seed coat. The gums are used in foods to act as a thickening or gelling agent.

Origin/Place of Manufacture

Southern USA, Central America, India, Pakistan.

Nutritional Benefit

High in protein (40%), with a reasonable amino acid profile. Average energy content with good level of minerals especially phosphorus and magnesium.

Colour/Texture

Dark brown/green meal/cake.

Palatability

Poor, due to its bitterness.

Limits to Usage (Anti-Nutritional Factors)

Trypsin inhibitors naturally present should be removed by processing, but the product should also be analysed before using. High in molybdenum, which can interfere with copper metabolism. Gums remaining limit its usage, especially in monogastrics.

Concentrate Inclusion % per species

	Inc %		Inc %		Inc %
Calf	0	Creep	0	Chick	0
Dairy	7.5	Weaner	0	Broiler	0
Beef	7.5	Grower	2.5	Breeder	2.5
Lamb	0	Finisher	5	Layer	2.5
Ewe	5	Sow	5		

Storage/Processing

Alternative Names

Bulk Density Meal 525 - 700 kg/m³

Typical Analysis

Dry Matter	92.0	NCDG	-	DUP (@ 5)	-	Avail Lysine	-
Crude Protein	40.0	NDF	31.9	DUP (@ 8)	-	Methionine	0.5
DCP	36.0	ADF	-	Salt	-	Meth & Cysteine	-
MER	12.1	Starch	3.0	Ca	0.4	Tryptophan	0.8
MEP	-	Sugar	3.0	Total Phos	0.6	Threonine	1.2
DE	11.6	Starch + Sugars	6.0	Av Phos	-	Arginine	0.5
Crude Fibre	12.0	FME	-	Magnesium	0.4	PDIA	-
Oil (EE)	5.0	ERDP (@ 2)	-	Potassium	-	PDIN	-
Oil (AH)	-	ERDP (@ 5)	-	Sodium	0.1	PDIE	-
EFA	-	ERDP (@ 8)	-	Chlorine	0.1	Met DI	-
Ash	5.1	DUP (@ 2)	-	Total Lysine	0.7	Lys DI	-

- Starch 3%
- Sugars 3%
- Protein 40%
- NDF 31.9%
- Ash 5.1%
- Oil 5%
- Other 12%

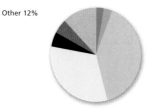

CONTEXT

Illipe Meal

Oilseeds and By-Products	

Introduction

A small brittle seed grown for its oil, which is crushed and removed by solvent extraction.

Origin/Manufacture

Asia.

Nutritional Benefit

Illipe meal is of low nutritional value, with protein ranging between 11 - 18%.

Colour/Texture

Reddish brown meal or pellet.

Palatability

Poor.

Limits to Usage (Anti-Nutritional Factors)

Contains a high tannin content which can reduce protein digestibility. Can also contain other toxic compounds.

Concentrate Inclusion % per species

	Inc %		Inc %		Inc %
Calf	0	Creep	0	Chick	0
Dairy	5.0	Weaner	0	Broiler	0
Beef	5.0	Grower	0	Breeder	0
Lamb	0	Finisher	0	Layer	0
Ewe	5.0	Sow	0		

Storage/Processing

Alternative Names

Bulk Density

Typical Analysis

Dry Matter	90.0	NCDG	-	DUP (@ 5)	-	Avail Lysine	-
Crude Protein	17.0	NDF	-	DUP (@ 8)	-	Methionine	-
DCP	12.0	ADF	-	Salt	-	Meth & Cysteine	-
MER	-	Starch	-	Ca	0.4	Tryptophan	-
MEP	-	Sugar	-	Total Phos	0.3	Threonine	-
DE	-	Starch + Sugars	-	Av Phos	0.1	Arginine	-
Crude Fibre	8.0	FME	-	Magnesium	0.3	PDIA	-
Oil (EE)	4.0	ERDP (@ 2)	-	Potassium	1.9	PDIN	-
Oil (AH)	4.3	ERDP (@ 5)	-	Sodium	0.1	PDIE	-
EFA	-	ERDP (@ 8)	-	Chlorine	0.2	Met DI	-
Ash	7.0	DUP (@ 2)	-	Total Lysine	-	Lys DI	-

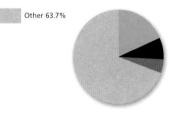

Starch -	NDF -	Other 63.7%	
Sugars -	Ash 7%		
Protein 17%	Oil 4.3%		

Forages and Stock Feeds

Introduction

Kale is an annual brassica, often grown to act as an extension to the grazing season in the autumn. Usually sown in June/July for feeding after harvesting. Sometimes it is ensiled as round bale kaleage.

Origin/Place of Manufacture

Europe.

Nutritional Benefit

Ideal for dairy cows but relatively low in dry matter (14%). It contains average protein levels (16-17%) and good levels of water soluble carbohydrates (20-25%). The feed value is proportional to the leaf content and mineral supplementation is essential, as the plant is high in calcium but low in phosphorus, manganese and iodine.
Cattle readily consume it fresh or ensiled.

Colour/Texture

Fleshy white grey tuber with green leaves.

Palatability

Good.

Limits to Usage (Anti-Nutritional Factors)

In common with all Brassica species the goitrogens must be taken into account particularly if other Brassicas are being fed eg rapeseed meal. Kale will also prevent thyroxine production if fed in excess amounts (greater than 20 kg/head/day for long periods).

High levels fed to dairy cows can result in a taint to the milk. Kale needs a good mineral supplement and particular attention should be taken over Iodine and Selenium levels.

Forage Inclusion % per species

Strip graze or 'cut and cart' at approximately 20 kg/head.
1 acre will last 100 cows 15 days (30 tonnes/acre).

	Inc %		Inc %		Inc %
Calf	10	Creep	0	Chick	0
Dairy	25	Weaner	0	Broiler	0
Beef	25	Grower	0	Breeder	0
Lamb	10	Finisher	0	Layer	0
Ewe	25	Sow	0		

Storage/Processing

Alternative Names

Bulk Density

Typical Analysis

Dry Matter	14.5	NDF	44.0	DUP (@ 8)	3.5	Methionine	-
Crude Protein	20.0	ADF	25.0	Salt	1.0	Meth & Cysteine	-
DCP	12.5	Starch	0.5	Ca	1.5	Tryptophan	-
MER	11.4	Sugar	17.0	Total Phos	0.4	Threonine	-
MEP	-	Starch + Sugars	17.5	Avail Phos	-	Arginine	-
DE	-	FME	9.0	Magnesium	0.2	PDIA	3.5
Crude Fibre	18.0	ERDP (@ 2)	15.0	Potassium	3.0	PDIN	11.5
Oil (EE)	2.1	ERDP (@ 5)	14.0	Sodium	0.1	PDIE	9.5
Oil (AH)	2.5	ERDP (@ 8)	13.0	Chloride	-	Met DI	-
Ash	15.0	DUP (@ 2)	1.8	Total Lysine	-	Lys DI	-
NCGD	86.0	DUP (@ 5)	2.8	Avail Lysine	-		

Starch 0.5% NDF 44% Other 1%

Sugars 17% Ash 15%

Protein 20% Oil 2.5%

Legumes and By-Products	

Introduction

Seeds of Lens culinaris a.o. Medik.

Grown for human food with substandard lentils and/or lentil bran available for animal feed.

Origin/Place of Manufacture

Asia, India, Eastern Europe.

Nutritional Benefit

Often considered close to beans or peas in analysis and also as a protein source. The bran is higher in fibre and lower in protein.

Colour/Texture

Orange/green seed.

Palatability

Good.

Limits to Usage (Anti-Nutritional Factors)

Can contain trypsin inhibitors and/or haemagglutinins at low level.

Concentrate Inclusion % per species

	Inc %		Inc %		Inc %
Calf	5	Creep	5	Chick	5
Dairy	10	Weaner	7.5	Broiler	5
Beef	12.5	Grower	10	Breeder	5
Lamb	5	Finisher	10	Layer	7.5
Ewe	10	Sow	12.5		

Storage/Processing

Stores well.

Alternative Names

Split peas, Red dahl.

Bulk Density

760 - 790 kg/m³

Typical Analysis

Dry Matter	88.0	NCDG	-	DUP (@ 5)	-	Avail Lysine	-
Crude Protein	29.0	NDF	-	DUP (@ 8)	-	Methionine	0.5
DCP	-	ADF	-	Salt	-	Meth & Cysteine	0.7
MER	4.0	Starch	-	Ca	-	Tryptophan	0.3
MEP	-	Sugar	-	Total Phos	0.3	Threonine	1.2
DE	-	Starch + Sugars	-	Av Phos	0.2	Arginine	2.5
Crude Fibre	4.0	FME	-	Magnesium	-	PDIA	-
Oil (EE)	1.7	ERDP (@ 2)	-	Potassium	-	PDIN	-
Oil (AH)	1.9	ERDP (@ 5)	-	Sodium	-	PDIE	-
EFA	-	ERDP (@ 8)	-	Chlorine	-	Met DI	-
Ash	3.0	DUP (@ 2)	-	Total Lysine	1.9	Lys DI	-

- Starch -
- Sugars -
- Protein 29%
- NDF -
- Ash 3%
- Oil 1.9%
- Other 62.1%

Miscellaneous

Introduction
Product obtained by grinding sources of calcium carbonate such as limestone, oyster or mussel shells, or by precipitation from acid solution.

Extracted from limestone quarries, this product is the most cost effective source of calcium available. The product varies in particle size from a flour to a grit.

Origin/Place of Manufacture
UK, Europe.

Nutritional Benefit
A major source of supplementary calcium.

Colour/Texture
Pale Grey.

Palatability
Not fed straight, so does not affect palatability.

Limits to Usage (Anti-Nutritional Factors).
May contain naturally occurring heavy metals with most suppliers providing guarantees.

Concentrate Inclusion % per species

	Inc %		Inc %		Inc %
Calf	1.5	Creep	1	Chick	1.5
Dairy	1.5	Weaner	1	Broiler	1.5
Beef	1.5	Grower	1	Breeder	1.5
Lamb	1.5	Finisher	1	Layer	7.0
Ewe	1.5	Sow	1.5		

Storage/Processing
Inert and stores well for long periods.

Alternative Names
Calcium Carbonate.

Bulk Density
Limestone Ground 1100 - 1450 kg/m³

Typical Analysis

Dry Matter	99.0	NCGD	0	ERDP (@ 5)	0	Sodium	0.05
Crude Protein	0	NDF	0	ERDP (@ 8)	0	Chloride	0
DCP	0	ADF	0	DUP (@ 2)	0	Total Lysine	0
MER	0	Cellulose	0	DUP (@ 5)	0	Methionine	0
MEP	0	Lignin	0	DUP (@ 8)	0	Meth & Cysteine	0
DE	0	Starch	0	Salt	0	Tryptophan	0
Crude Fibre	0	Sugar	0	Ca	38.0	Threonine	0
Oil (EE)	0	Starch + Sugars	0	Total Phos	0	Arginine	0
Oil (AH)	0	FME	0	Magnesium	0.15	PDIA	0
Ash	97.0	ERDP (@ 2)0	0	Potassium	0		

Starch 0%   NDF 0%   Other 0%

Sugars 0%   Ash 100%

Protein 0%   Oil 0%

Oilseeds and By-Products

Introduction

Seeds of linseed Linum usitatissimum L. (Minimum botanical purity 93%).

Linseed is the seed of the flax plant and is grown for its oil which is used in paints, inks, soaps etc. Oil is either expelled or extracted from the seed which is made up of approximately 39% oil. The meal can then be pressed into lozenge shaped pellets or chips for easy transportation. The whole seed is not widely used.

Origin/Place of Manufacture

Europe, America, Argentina and China.

Nutritional Benefit

It is high in energy, oil, available carbohydrates, and of mid-protein, but low in limiting amino acids. The oil present is high in polyunsaturated fatty acids eg. linolenic acid (C18:3). Linseed is reported to aid bloom in coat condition and, with large amounts fed, milk fat content can be reduced, but so too may overall digestibility of the ration. It is a good source of digestible fibre and a palatable feed. High in phosphorus but up to two-thirds can be in the form of phytate. It is low in calcium and lysine.

Whole linseed has been fed on farm after treatment with caustic soda and water. It is claimed that this whole oilseed, coated in sodium bicarbonate ensures oil digestion bypasses the rumen and takes place in the intestine.

Colour/Texture

Tan - brown meal, cake, lozenge or pellets.

Palatability

Good.

Limits to Usage (Anti-Nutritional Factors)

May contain glucoside or linamarin and the enzyme linase, but good processing at high temperatures will remove most. High levels will have a laxative effect, and result in soft carcass fat. It is prone to oxidation; also can lead through to oxidation of fat in animal milk. Contains 5-10% mucilage which can be digested by ruminants, but is not ideal for poultry due to adverse effect on B Vitamins and will retard growth above 5% inclusion.

Concentrate Inclusion % per species - Linseed Meal

	Inc %		Inc %		Inc %
Calf	7.5	Creep	0	Chick	0
Dairy	20	Weaner	0	Broiler	0
Beef	20	Grower	0	Breeder	2.5
Lamb	7.5	Finisher	10	Layer	2.5
Ewe	20	Sow	5		

Concentrate Inclusion % per species - Linseed Whole

	Inc %		Inc %		Inc %
Calf	0	Creep	0	Chick	0
Dairy	5	Weaner	0	Broiler	0
Beef	5	Grower	0	Breeder	0
Lamb	0	Finisher	0	Layer	0
Ewe	5	Sow	0		

Storage/Processing

Often processed to form lozenges (small cakes) and to destroy toxic compounds.

Alternative Names

Flax seed expeller meal.

Bulk Density

Meal 560 - 650 kg/m³   **Cake** 700 - 800 kg/m³

Typical Analysis	Linseed Whole	Expellers	Extraction
Dry Matter	90.0	90.0	88.0
Crude Protein	23.0	36.0	38.0
DCP	18.5	31.0	33.0
MER	18.9	12.8	13.2
MEP	20.5	11.5	11.5
DE	20.5	14.9	14.9
Crude Fibre	7.0	10.5	11.5
Oil (EE)	37.0	8.1	7.1
Oil (AH)	39.0	8.5	7.1
EFA	29.5	5.0	5.0
Ash	4.5	7.0	6.5
NCGD	90.0	85.0	85.0
NDF	28.0	18.9	25.1
ADF	8.5	10.1	15.0
Starch	10.0	4.5	8.5
Sugar	3.5	4.5	8.5
Starch + Sugars	13.5	9.0	13.0
FME	5.5	10.0	10.7
ERDP (@ 2)	18.5	24.5	25.5
ERDP (@ 5)	15.5	20.1	21.2
ERDP (@ 8)	13.5	17.5	18.6
DUP (@ 2)	4.5	6.2	6.7
DUP (@ 5)	7.2	9.2	9.7
DUP (@ 8)	9.3	12.9	13.6
Salt	0.1	0.1	0.3
Ca	0.2	1.0	0.5
Total Phos	0.7	1.0	1.0
Av Phos	0.03	0.4	0.4
Magnesium	0.9	0.6	0.6
Potassium	0.7	1.2	1.0
Sodium	0.1	0.1	0.1
Chloride	0.1	0.05	0.05
Total Lysine	0.9	1.3	1.3
Avail Lysine	0.6	1.0	1.0
Methionine	0.45	0.7	0.7
Meth & Cysteine	0.75	1.3	1.3
Tryptophan	0.3	0.6	0.6
Threonine	0.8	1.2	1.2
Arginine	2.0	3.0	3.0
PDIA	3.4	12.5	12.5
PDIN	14.3	23.2	23.2
PDIE	6.5	16.9	16.9
Met DI	0.15	0.3	0.3
Lys DI	0.45	0.6	0.6

- Starch 8.5%
- Sugars 8.5%
- Protein 38%
- NDF 25.1%
- Ash 6.5%
- Oil 7.1%
- Other 6.3%

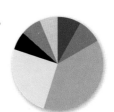

Locust Bean

Roots, Fruits and By-Products

Introduction

Product obtained by crushing the dried fruits (Pods) of the carob tree Ceratonia Siliqua L. from which the locust beans have been removed.

Produced from the carob tree when the bean is ripe, the flesh of the seed is removed to leave a seed. It is either used for animal feed or for alcohol production and may be ground with the pod.

Origin/Place of Manufacture

Mediterranean countries and sub-tropical regions.

Nutritional Benefit

High in sugars (+40%), it is extremely palatable and usually added to coarse mixes to improve palatability. It is often poorly digested. It is commonly used in coarse mixes, probably for its appearance and palatability.

Colour/Texture

Dark brown, usually in flake form.

Palatability

Good.

Limits to Usage (Anti-Nutritional Factors)

Part of its protein may be in the form of tannins (tannic acid) reducing digestibility. Rich in sugar and can, therefore, increase risk of acidosis in ruminants.

Concentrate Inclusion % per species

	Inc %		Inc %		Inc %
Calf	0	Creep	0	Chick	0
Dairy	10	Weaner	0	Broiler	2.5
Beef	12.5	Grower	2.5	Breeder	5
Lamb	5	Finisher	5.0	Layer	5
Ewe	10	Sow	7.5		

Storage/Processing

Difficult to grind due to high sugar levels.

Alternative Names

Kibbled Beans, Carob Pods, Ribbled Locust Beans.

Bulk Density

Typical Analysis

Dry Matter	90.0	NCGD	81.0	DUP (@ 5)	0.7	Avail Lysine	0.3
Crude Protein	6.0	NDF	50.1	DUP (@ 8)	0.9	Methionine	0.08
DCP	3.5	ADF	42.4	Salt	0.35	Meth & Cysteine	0.15
MER	11.9	Starch	1.0	Ca	0.45	Tryptophan	0.05
MEP	8.9	Sugar	34.0	Total Phos	0.2	Threonine	0.2
DE	10.6	Starch + Sugars	35.0	Av Phos	0.1	Arginine	0.34
Crude Fibre	8.0	FME	10.0	Magnesium	0.05	PDIA	5.9
Oil (EE)	1.1	ERDP (@ 2)	3.6	Potassium	0.60	PDIN	33.0
Oil (AH)	1.4	ERDP (@ 5)	3.3	Sodium	0.02	PDIE	68.0
EFA	0.3	ERDP (@ 8)	3.1	Chlorine	0.2	Met DI	0.04
Ash	3.9	DUP (@ 2)	0.5	Total Lysine	0.35	Lys DI	0.3

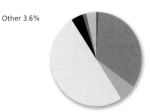

Starch 1%

NDF 50.1%

Other 3.6%

Sugars 34%

Ash 3.9%

Protein 6%

Oil 1.4%

Forages and Stock Feeds		

Introduction

Product obtained by drying and milling young lucerne Medicago savita L. and Medicago var. Martyn (minimum botanical purity 80%).

A perennial legume with many leaves. It is dried, milled and pelleted (a little molasses may be added to aid pelleting). It is known in the USA as Alfalfa.

Origin/Place of Manufacture

USA, Europe.

Nutritional Benefit

A good protein and digestible fibre source for ruminants. Similar, if not higher in quality, to grass cubes, with good protein and vitamin supply. The analysis varies, depending upon the stage of harvest and processing conditions. High in digestible fibre, Undegradable Protein, some vitamins and calcium. Good source of beta carotene.

Colour/Texture

Green hard pellets.

Palatability

Good.

Limits to Usage (Anti-Nutritional Factors)

High in NDF, but in the form of pellets or meal does not act like a normal 'long' forage. Can be high in potassium and this needs to be noted in high quality grass silage diets using molasses.

Concentrate Inclusion % per species

	Inc %		Inc %		Inc %
Calf	20	Creep	0	Chick	0
Dairy	30	Weaner	0	Broiler	0
Beef	30	Grower	0	Breeder	0
Lamb	20	Finisher	0	Layer	2.5
Ewe	30	Sow	0.5		

Storage/Processing

Alternative Names

Alfalfa.

Bulk Density

Meal 325 - 375 kg/m³   **Pellets** 600 - 650 Kg/m³

Typical Analysis

Dry Matter	90.0	NCGD	61.0	DUP (@ 5)	3.5	Avail Lysine	0.45
Crude Protein	18.0	NDF	55.0	DUP (@ 8)	4.1	Methionine	0.25
DCP	12.0	ADF	33.0	Salt	1.0	Meth & Cysteine	0.45
MER	10.0	Starch	2.0	Ca	1.5	Tryptophan	0.3
MEP	6.0	Sugar	6.5	Total Phos	0.3	Threonine	0.6
DE	7.5	Starch + Sugars	8.5	Av Phos	0.15	Arginine	0.7
Crude Fibre	23.0	FME	7.3	Magnesium	0.25	PDIA	6.1
Oil (EE)	3.0	ERDP (@ 2)	12.5	Potassium	2.5	PDIN	12.6
Oil (AH)	3.5	ERDP (@ 5)	11.1	Sodium	0.15	PDIE	9.8
EFA	0.5	ERDP (@ 8)	11.3	Chloride	0.5	Met DI	0.18
Ash	10	DUP (@ 2)	2.1	Total Lysine	0.75	Lys DI	0.63

Starch 2%   NDF 55%   Other 5%

Sugars 6.5%   Ash 10%

Protein 18%   Oil 3.5%

Forages and Stock Feeds	

Introduction
Lucerne, a deep rooted legume, is now being grown as a forage crop in many areas where shallow soil or poor rainfall produces poor summer grass growth. Lucerne can be grown as a straight crop or undersown with spring barley or forage maize.

Origin/Place of Manufacture
Mid and Southern UK, Europe.

Nutritional Benefit
The crop is usually wilted to 25-30% dry matter. It is a higher protein content (19% in DM) than grass silage. The silage is claimed to be high in Undegradable Protein compared to grass silage, and also to result in higher intake levels due to lower cell wall content. Cattle readily consume well made lucerne silage.

Colour/Texture
Dark green silage.

Palatability
Good.

Limits to Usage
High fibre levels may reduce nutrient density in high yielding rations.

Forage Inclusion % per species

	Inc%		Inc%		Inc%
Calf	100	Creep	0	Chick	0
Dairy	100	Weaner	0	Broiler	0
Beef	100	Grower	0	Breeder	0
Lamb	100	Finisher	0	Layer	0
Ewe	100	Sow	0		

Storage/Processing

Alternative Names
Alfalfa silage

Bulk Density
600 - 650 kg/m³ after consolidation.

Typical Analysis

Dry Matter	35.0	NCGD	-	DUP (@ 5)	-	Avail Lysine	-
Crude Protein	19.5	NDF	62.0	DUP (@ 8)	0	Methionine	0.25
DCP	17.5	ADF	37.2	Salt	0.05	Meth & Cysteine	0.45
MER	8.5	Starch	0.5	Ca	1.5	Tryptophan	0.2
MEP	-	Sugar	1.0	Total Phos	0.3	Threonine	0.7
DE		Starch + Sugars	1.5	Av Phos	-	Arginine	0.5
Crude Fibre	30.0	FME	7.3	Magnesium	0.2	PDIA	-
Oil (EE)	7.0	ERDP (@ 2)	-	Potassium	0	PDIN	-
Oil (AH)	7.0	ERDP (@ 5)	-	Sodium	0.02	PDIE	-
EFA	-	ERDP (@ 8)	13.0	Chloride	0	Met DI	-
Ash	10.0	DUP (@ 2)	-	Total Lysine	0.8	Lys DI	-

Starch 0.5% NDF 62% Other 0%
Sugars 1% Ash 10%
Protein 19.5% Oil 7%

CONTEXT

Legumes and By-Products

Introduction

Seeds of Lupinus spp. low in bitter seed content.

Grown as a break crop, lupins produce small yellow/brown peas which are flaked or milled. There are three flowering types: white, yellow and blue. Lupin seed meal is often an economic alternative to soybean meal and rapeseed meal. Amino acid supplementation along with enzyme addition and processing have removed any performance reducing effects of the past and now low alkaloid sweet varieties are available.

Origin/Place of Manufacture

Australia, Europe, S. E. Asia.

Nutritional Benefit

Good protein source (32%). It is essential that the lupin peas are processed prior to feeding to remove the fibrous seed coat which, if allowed to remain in feed, will affect the chemical analysis. Mostly white and yellow flowering varieties are used to make the meal. NB. Methionine supplementation may be required and lysine availability raised across species.

Colour/Texture

Cream/grey flakes or meal.

Palatability

There are both sweet and unpalatable varieties.

Limits to Usage (Anti-Nutritional Factors)

Alkaloids are present in bitter varieties and are toxic but this is generally not a problem in new varieties. High fibre level makes it poorly digested by young monogastrics. Low in Methionine and even Lysine.

Concentrate Inclusion % per species

	Inc %		Inc %		Inc %
Calf	7.5	Creep	0	Chick	0
Dairy	12.5	Weaner	0	Broiler	5
Beef	15	Grower	0	Breeder	7.5
Lamb	2.5	Finisher	7.5	Layer	7.5
Ewe	12.5	Sow	7.5		

Storage/Processing

Must be rolled, ground or flaked prior to feeding. Enzyme addition may also be required to break down non-starch polysaccharides. High oil level will mean they may go rancid quickly after milling. Heat processing may allow high intake levels. Soaking in water can improve performance in broilers.

Alternative Names

Sweet Lupins.

Bulk Density

Meal 325 - 375 kg/m³

Typical Analysis

Dry Matter	86.0	NCDG	81.0	DUP (@ 5)	5.7	Avail Lysine	1.0
Crude Protein	32.0	NDF	33.1	DUP (@ 8)	8.1	Methionine	0.3
DCP	28.0	ADF	15.9	Salt	0.2	Meth & Cysteine	0.83
MER	14.5	Starch	9.0	Ca	0.4	Tryptophan	0.3
MEP	11.0	Sugar	4.0	Total Phos	0.4	Threonine	1.2
DE	13.2	Starch + Sugars	13.0	Av Phos	0.2	Arginine	3.5
Crude Fibre	13.5	FME	8.0	Magnesium	0.55	PDIA	1.4
Oil (EE)	5.3	ERDP (@ 2)	30.2	Potassium	1.0	PDIN	21.6
Oil (AH)	6.2	ERDP (@ 5)	26.0	Sodium	0.05	PDIE	6.9
EFA	5.2	ERDP (@ 8)	24.2	Chlorine	0.1	Met DI	0.15
Ash	3.7	DUP (@ 2)	2.3	Total Lysine	1.9	Lys DI	0.47

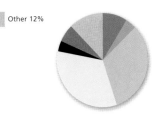

Starch 9%	NDF 33.1%	Other 12%	
Sugars 4%	Ash 3.7%		
Protein 32%	Oil 6.2%		

Cereals and By-Products

Introduction

By-product of oil manufacture, obtained by pressing/extraction of dry or wet maize germ to which parts of the endosperm and testa may still adhere.

As the name implies, these are the dried germs from maize after the oil has been extracted. Maize germ meal is a by-product from either the wet milling of maize, produced after the initial grinding and separation, or from the production of corn, often in the production of cornflakes and lager. When wet milling maize, the germs (being 50% oil) are removed early in the process. The oil is either expelled or extracted and the removal rate will determine what is left in the meal. There are two main types of maize germ meal: high (14%+) and low oil (5%).

In some production plants, the maize germ meal is added back to the maize fibre/maize gluten feed.

Origin/Place of Manufacture

UK, USA.

Nutritional Benefit

Quality depends on manufacturing process but generally ideal for all classes of livestock. The amount of bran remaining and residual oil level can affect suitability for ruminant and non-ruminant feeds.

Quite high in starch, energy and protein with good amino acid balance.
High oil products are not suitable for ruminant feed. The oil is free unsaturated and could be detrimental to rumen function if not formulated correctly. The low calcium content makes it suitable for feeding to dry cows.

Colour/Texture

Pale yellow/tan oily meal.

Palatability

Highly palatable.

Limits to Usage (Anti-Nutritional Factors)

High oil maize germ can produce soft fat in pigs and reduce fibre digestion in ruminants if fed for long periods. It can also cause scouring. Lower oil products have few limits to usage.

Concentrate Inclusion % per species

	Inc %		Inc %		Inc %
Calf	20	Creep	10	Chick	15
Dairy	25	Weaner	15	Broiler	20
Beef	25	Grower	15	Breeder	15
Lamb	15	Finisher	10	Layer	10
Ewe	20	Sow	15		

Storage/Processing

Will go mouldy and rancid if stored for long periods. The oil present, can oxidise leading to active peroxides in the feed which can reduce vitamin E levels. It is also prone to heating when stored for long periods.

Alternative Names

Bulk Density

Typical Analysis	Low Oil	High Oil
Dry Matter	88.0	88.0
Crude Protein	26.0	23.0
DCP	18.5	17.0
MER	14.3	15.9
MEP	11.3	12.3
DE	14.8	15.3
Crude Fibre	10.0	10.0
Oil (EE)	4.0	11.0
Oil (AH)	4.5	11.3
EFA	3.0	7.0
Ash	3.0	3.0
NCGD	80.3	76.0
NDF	36.5	33.0
ADF	14.1	13.5
Starch	23.0	21.0
Sugar	7.0	6.5
Starch + Sugars	30.0	27.5
FME	12.0	12.0
ERDP (@ 2)	19.3	18.8
ERDP (@ 5)	12.2	11.7
ERDP (@ 8)	15.1	10.5
DUP (@ 2)	1.9	1.8

Typical Analysis	Low Oil	High Oil
DUP (@ 5)	4.6	4.5
DUP (@ 8)	5.3	5.2
Salt	0.15	0.13
Ca	0.1	0.1
Total Phos	0.8	0.75
Av Phos	0.5	0.5
Magnesium	0.15	0.15
Potassium	0.3	0.3
Sodium	0.05	.0.03
Chloride	0.08	0.08
Total Lysine	1.2	1.15
Avail Lysine	0.6	0.6
Methionine	0.45	0.43
Meth & Cysteine	0.90	0.83
Tryptophan	0.25	0.25
Threonine	1.1	1.0
Arginine	2.1	2.0
PDIA	10.2	10.0
PDIN	16.3	16.0
PDIE	15.8	15.5
Met DI	0.3	0.3
Lys DI	1.0	1.0

Starch 23%   NDF 36.5%   Other 0%

Sugars 7%   Ash 3%

Protein 26%   Oil 4.5%

Cereals and By-Products

Introduction

Dried by-product of the manufacture of maize starch. It is composed of bran and gluten to which components of the steeping liquor, and possibly the germ, from which the oil may have been removed, may be added.

Derived from the wet milling of maize to extract starch or for alcohol/ethanol production, this feed consists of maize fibre and corn steep liquor, dried and pelleted. The product is manufactured mainly in USA but home-produced sources are also available.

The maize seed is rich in starch 60-80%, being found mainly in the endosperm. The grain is then usually steeped with a mild sulphurous acid solution for 36-48 hours. This softens the kernel and releases soluble protein. The steep liquor is removed to make corn steep liquor, the fibre is pelleted with it, to make maize gluten. The germ is floated off, washed and squeezed to remove moisture. To get at this starch, the fibre or bran is removed and the gluten meal removed. Maize gluten feed is generally a South America mixture of the fibre, corn steep liquor and potentially maize germ.

Origin/Place of Manufacture

USA, S. America, UK.

Nutritional Benefit

As a 20% protein feed with reasonable energy, maize gluten suits many rations and is usually of good digestibility. It contains good levels of digestible fibre starch (+20% in DM).

If the product is black, it may have been overheated in processing and should be avoided. The protein quality means it is of less use for pigs and poultry. Although it has reasonable fibre levels, due to milling, it is quite short in length and quickly fermented. Approximately 15% of the starch is rumen fermented. To avoid differences in analysis always buy from the same source.

Colour/Texture

This is a light brown, dried meal or pellet.

Palatability

Good.

Limits to Usage

The starch can be highly degradable. Protein levels can vary from source to source. Needs a good source of supplementary minerals when used at maximum levels for ruminant diets.

Concentrate Inclusion % per species

	Inc %		Inc %		Inc %
Calf	15	Creep	0	Chick	0
Dairy	35	Weaner	0	Broiler	12.5
Beef	40	Grower	10	Breeder	0
Lamb	30	Finisher	10	Layer	12.5
Ewe	35	Sow	10		

Storage/Processing

Alternative Names Corn Gluten Feed.

Bulk Density 600 - 650 kg/m³

Typical Analysis

Dry Matter	88.0	NCGD	73.2	DUP (@ 5)	2.7	Avail Lysine	0.4
Crude Protein	21.5	NDF	42.5	DUP (@ 8)	3.5	Methionine	0.5
DCP	17.0	ADF	9.9	Salt	0.65	Meth & Cysteine	0.9
MER	12.9	Starch	22.0	Ca	0.3	Tryptophan	0.15
MEP	9.1	Sugar	3.5	Total Phos	1.0	Threonine	0.85
DE	12.5	Starch + Sugars	25.5	Av Phos	0.5	Arginine	1.1
Crude Fibre	8.0	FME	11.8	Magnesium	0.45	PDIA	6.3
Oil (EE)	3.1	ERDP (@ 2)	17.2	Potassium	1.5	PDIN	14.8
Oil (AH)	4.0	ERDP (@ 5)	15.7	Sodium	0.25	PDIE	12.2
EFA	1.65	ERDP (@ 8)	14.8	Chloride	0.25	Met DI	0.25
Ash	6.5	DUP (@ 2)	1.27	Total Lysine	0.7	Lys DI	0.65

Starch 22%	NDF 42.5%	Other 0%
Sugars 3.5%	Ash 6.5%	
Protein 21.5%	Oil 4%	

Cereals and By-Products

Introduction

Dried by-product of the manufacture of maize starch. It consists principally of gluten obtained during the separation of the starch.

In the wet milling process, after the maize germ and fibre are removed, the remaining material is centrifuged to isolate the starch from the gluten for further processing to modified starches, sweeteners, etc. The remaining gluten is dried and then milled/sieved to produce a consistent coarse powder high in proteins with a natural pigment.

Origin/Place of Manufacture

USA, UK, Europe.

Nutritional Benefit

Rich in protein and energy but poor in lysine. It is also a reasonable starch source (16% in DM).

For ruminants, the protein is largely rumen undegradable and high in bypass methionine, while layer feeds benefit from the natural pigments which colour yolks.

Colour/Texture

Bright golden orange granular meal.

Palatability

Palatable up to 15% inclusion within a ration.

Limits to Usage (Anti-Nutritional Factors)

Excessive levels may cause soft carcass fat or discolouration.

Concentrate Inclusion % per species

	Inc %		Inc %		Inc %
Calf	5	Creep	0	Chick	0
Dairy	10	Weaner	2.5	Broiler	5
Beef	10	Grower	2.5	Breeder	0
Lamb	5	Finisher	5.0	Layer	7.5
Ewe	10	Sow	7.5		

Storage/Processing

Average storage qualities.

Alternative Names

Prairie Meal, Gluten 60 meal.

Bulk Density

550 - 650 kg/m³

Typical Analysis

Dry Matter	90.0	NCGD	95.0	DUP (@ 5)	35.5	Avail Lysine	1.4
Crude Protein	68.0	NDF	7.5	DUP (@ 8)	40.0	Methionine	1.8
DCP	68.0	ADF	11.5	Salt	0.1	Meth & Cysteine	3.3
MER	14.5	Starch	15.5	Ca	0.01	Tryptophan	0.35
MEP	17.7	Sugar	1.0	Total Phos	0.5	Threonine	2.5
DE	18.7	Starch + Sugars	16.5	Av Phos	0.15	Arginine	2.4
Crude Fibre	1.5	FME	12.5	Magnesium	0.1	PDIA	50.5
Oil (EE)	4.5	ERDP (@ 2)	35.5	Potassium	0.15	PDIN	58.0
Oil (AH)	6.0	ERDP (@ 5)	23.5	Sodium	0.1	PDIE	53.0
EFA	1.5	ERDP (@ 8)	18.5	Chloride	0.05	Met DI	1.4
Ash	2.0	DUP (@ 2)	24.5	Total Lysine	1.5	Lys DI	2.0

- Starch 15.5%
- NDF 7.5%
- Other 0%
- Sugars 1%
- Ash 2%
- Protein 68%
- Oil 6%

Cereals and By-Products

Introduction

Grains of Zea Mays L.

Maize is grown as a food ingredient, as a substrate for fermentation, milling or for animal feed. There are three types: white, red and yellow, with the latter the main type used for animal feed. It is not usually used straight in the UK and Europe as it is expensive.

Origin/Place of Manufacture

Temperate countries, with the largest volumes grown in the USA, France and South America.

Nutritional Benefit

The highest energy of most cereals, containing twice the oil of wheat and barley but lower and poorer in protein quality than wheat. The seed is high in starch (65 - 70%) but low in protein (10.5%), fibre and minerals. The starch is slowly degraded compared to wheat and barley with reports of it being partially resistant to rumen degradation; 30% of the starch is rumen unfermented. Low in calcium and usually only added to pig and poultry feeds. Flaked form used in coarse feeds. Its low minerals status needs careful supplementation.

Colour/Texture

Large yellow/orange pea-like coarse grain or coarse golden flake.

Palatability Good.

Limits to Usage (Anti-Nutritional Factors)

It has been reported that at high levels, when fed for prolonged periods, yellow carcass fat results due to the xanthophyll content.

Concentrate Inclusion % per species

	Inc %		Inc %		Inc %
Calf	35	Creep	0	Chick	30
Dairy	35	Weaner	50	Broiler	50
Beef	35	Grower	40	Breeder	50
Lamb	35	Finisher	25	Layer	50
Ewe	35	Sow	35		

Storage/Processing

The seed should be flaked, cracked, rolled, micronised, jet sploded or ground before feeding depending on the animal it is to be fed to. The processing will improve the digestibility of the grain.

Alternative Names

Indian Corn, Corn, Yellow Maize.

Bulk Density

725 - 775 kg/m³

Typical Analysis

Dry Matter	89.0	NCGD	93.0	DUP (@ 5)	1.7	Avail Lysine	0.5
Crude Protein	9.6	NDF	14.5	DUP (@ 8)	2.3	Methionine	0.2
DCP	8.5	ADF	2.6	Salt	0.1	Meth & Cysteine	0.45
MER	14.5	Starch	68.0	Ca	0.1	Tryptophan	0.1
MEP	16.0	Sugar	2.0	Total Phos	0.3	Threonine	0.6
DE	17.1	Starch + Sugars	75.0	Av Phos	0.15	Arginine	0.5
Fibre	2.5	FME	12.8	Magnesium	0.1	PDIA	6.0
Oil (EE)	4.1	ERDP (@ 2)	8.0	Potassium	0.4	PDIN	8.0
Oil (AH)	4.4	ERDP (@ 5)	7.2	Sodium	0.1	PDIE	13.0
EFA	2.3	ERDP (@ 8)	7.0	Chloride	0.1	Met DI	0.28
Ash	1.5	DUP (@ 2)	1.0	Total Lysine	0.8	Lys DI	0.8

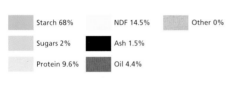

Starch 68%   NDF 14.5%   Other 0%

Sugars 2%   Ash 1.5%

Protein 9.6%   Oil 4.4%

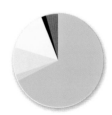

Cereals and By-Products	

Introduction

Maize for wet milling is screened to remove broken and undersized kernels and flour.
A whole kernel is usually required for the wet milling process, and 5% maize screenings
are, therefore, usually produced. Maize screenings are an economic alternative to
whole maize and even to wheat/barley as they are already processed and do not
require grinding.

Origin/Manufacture

UK.

Nutritional Benefit

The product is high in starch (60-80%) with good energy levels. It has been reported to
be high in rumen resistant starch and to be ideal to enhance milk protein levels.
Approximately 30% of the starch is unfermented in the rumen.

Colour/Texture

Pale yellow/white coarse meal.

Palatability

Good.

Limits to Usage (Anti-Nutritional Factors)

Occasionally, whole grains may appear in animal faeces; this is generally not a problem
and wastage is small.

Concentrate Inclusion % per species

	Inc %		Inc %		Inc %
Calf	10	Creep	10	Chick	10
Dairy	30	Weaner	15	Broiler	20
Beef	30	Grower	25	Breeder	25
Lamb	15	Finisher	25	Layer	30
Ewe	20	Sow	25		

Storage/Processing

Alternative Names

Corn Meal.

Bulk Density

480 Kg/m³

Typical Analysis

Dry Matter	90.0	NCGD	88.0	DUP (@ 5)	1.7	Avail Lysine	0.6
Crude Protein	9.5	NDF	16.7	DUP (@ 8)	2.3	Methionine	2.1
DCP	9.0	ADF	2.6	Salt	0.1	Meth & Cysteine	0.45
MER	14.5	Starch	68.0	Ca	0.1	Tryptophan	0.1
MEP	14.8	Sugar	1.0	Total Phos	0.3	Threonine	0.6
DE	16.5	Starch + Sugars	69.0	Av Phos	0.15	Arginine	0.5
Crude Fibre	3.5	FME	12.7	Magnesium	0.1	PDIA	6.0
Oil (EE)	2.5	ERDP (@ 2)	8.1	Potassium	0.4	PDIN	8.0
Oil (AH)	3.8	ERDP (@ 5)	7.2	Sodium	0.1	PDIE	13.0
EFA	2.0	ERDP (@ 8)	7.0	Chloride	0.1	Met DI	0.3
Ash	1.0	DUP (@ 2)	1.0	Total Lysine	1.0	Lys DI	0.8

Starch 68%   NDF 16.7%   Other 0%

Sugars 1%   Ash 1%

Protein 9.5%   Oil 3.8%

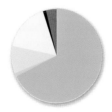

Forages and Stock Feeds

Introduction

Growing in popularity as a higher energy forage, it produces excellent Dry Matter yields (10 tonnes of DM/ha) with only one harvest. It also leaves ground ideal for the application of farm yard manure in the early spring which has been accumulating on farm throughout the winter.

Yields, and the ability to grow the crop across most of the UK, have improved dramatically over recent years.

Origin/Place of Manufacture

Grown in France, Holland etc. and now in most of England, parts of Wales and even Scotland.

Nutritional Benefit

High in starch (+25%) but low in protein with a good Dry Matter level (+30%). Ideal in a complete mix and has been shown to increase overall forage intakes. It is naturally low in mineral and trace element levels. Maize silage and other forages mixed together will stimulate Dry Matter Intake.

Colour/Texture Brown/yellow/green fibrous material.

Palatability

Good.

Limits to Usage (Anti-Nutritional Factors)

Grains must be damaged or cracked at harvest or they will pass through the animal undigested especially at higher inclusion rates. Protein supplements usually required. The short chop length needed for good consolidation means it is limited in 'structural fibre' for dairy cows.

The nutritive value varies dramatically so appropriate analysis should be undertaken.

Inclusion % of forage per species

Can be fed up to 99% of forage but 75% is the usual maximum.

	Inc %		Inc %		Inc %
Calf	75	Creep	0	Chick	0
Dairy	75	Weaner	0	Broiler	0
Beef	75	Grower	0	Breeder	0
Lamb	75	Finisher	0	Layer	0
Ewe	75	Sow	0		

Storage/Processing

Stores well if sealed properly. Rock salt is often added to the surface to reduce spoilage.

Alternative Names

Corn silage.

Bulk Density

650 - 750 kg/m³ in well consolidated clamps.

Typical Analysis

Dry Matter	30.0	NCGD	71.0	DUP (@ 5)	1.6	Avail Lysine	-
Crude Protein	9.0	NDF	55.0	DUP (@ 8)	1.9	Methionine	0.12
DCP	6.0	ADF	30.0	Salt	0.5	Meth & Cysteine	0.25
MER	11.5	Starch	22.0	Ca	0.2	Tryptophan	0.05
MEP	-	Sugar	3.0	Total Phos	0.2	Threonine	0.2
DE	-	Starch + Sugars	25.0	Av Phos	-	Arginine	0.2
Crude Fibre	18	FME	10.1	Magnesium	0.1	PDIA	2.0
Oil (EE)	3.0	ERDP (@ 2)	7.0	Potassium	0.6	PDIN	6.0
Oil (AH)	3.5	ERDP (@ 5)	6.7	Sodium	0.2	PDIE	7.0
EFA	-	ERDP (@ 8)	6.5	Chloride	0.2	Met DI	0.1
Ash	4.5	DUP (@ 2)	1.3	Total Lysine	0.2	Lys DI	0.2

Starch 22%		NDF 55%	Other 3%
Sugars 3%		Ash 4.5%	
Protein 9%		Oil 3.5%	

Cereals and By-Products

Introduction

By-product of malting, consisting mainly of dried rootlets of germinated cereals.

These are the dried rootlets and sprouts usually from germinated barley that have occurred during the malting process.

Place of Manufacture

UK, USA.

Nutritional Benefit

A good source of protein (25%) of average digestibility and average energy level assuming they have not been over dried. Overdrying will reduce the protein digestibility of the grain. Generally low in starch but high in digestible fibre, digestible protein and some sugars (13% in DM).

Can usefully be included to a small extent in sheep diets. They make a useful contribution to the protein content of the diet but are not usually available in large quantities.

Colour/Texture

Brown/tan meal or pellets.

Palatability

Slightly bitter.

Limits to Usage (Anti-Nutritional Factors)

High fibre and moderate energy content restricts usage to ruminant rations. They absorb moisture easily, causing them to swell which can cause problems in the rumen when fed at high levels.

Concentrate Inclusion % per species

	Inc %		Inc %		Inc %
Calf	0	Creep	0	Chick	0
Dairy	15	Weaner	0	Broiler	0
Beef	15	Grower	3	Breeder	5
Lamb	2.5	Finisher	5.0	Layer	5
Ewe	7.5	Sow	7.0		

Storage/Processing

Very fluffy.

Alternative Names

Malt Sprouts.

Bulk Density

600Kg/m³

Typical Analysis

Dry Matter	90.0	NCGD	38.0	DUP (@ 5)	2.7	Avail Lysine	0.8
Crude Protein	22.5	NDF	52.7	DUP (@ 8)	3.8	Methionine	0.45
DCP	16.0	ADF	16.5	Salt	1.1	Meth & Cysteine	1.1
MER	11.5	Starch	6.0	Ca	0.3	Tryptophan	0.2
MEP	7.5	Sugar	11.0	Total Phos	0.7	Threonine	1.0
DE	10.5	Starch + Sugars	17.0	Av Phos	0.25	Arginine	1.2
Crude Fibre	13.5	FME	10.5	Magnesium	0.2	PDIA	5.1
Oil (EE)	2.5	ERDP (@ 2)	21.0	Potassium	1.7	PDIN	16.1
Oil (AH)	2.5	ERDP (@ 5)	19.5	Sodium	0.05	PDIE	10.7
EFA	0.6	ERDP (@ 8)	18.5	Chloride	0.6	Met DI	-
Ash	5.3	DUP (@ 2)	1.4	Total Lysine	1.3	Lys DI	-

Starch 6%	NDF 52.7%	Other 0%
Sugars 11%	Ash 5.3%	
Protein 22.5%	Oil 2.5%	

Malt Residual Pellets

Cereals and By-Products	

Introduction
Made from the chitted barley roots (malt culms) and malt screenings from the malting process.

Origin/Place of Manufacture
Mainly produced in the UK.

Nutritional Benefit
Average energy and often used as a lower cost filler, although a reasonable source of protein.

Colour/Texture
An orange/brown pellet with yellow/white speck.

Palatability
Average.

Limits to Usage (Anti-Nutritional Factors)
High fibre content may restrict usage to ruminant diets of lower productivity. They absorb moisture easily, causing them to swell and can cause problems in the rumen.

Concentrate Inclusion Rate % per species

	Inc %		Inc %		Inc %
Calf	10	Creep	0	Chick	0
Dairy	20	Weaner	0	Broiler	0
Beef	25	Grower	0	Breeder	0
Lamb	10	Finisher	0	Layer	0
Ewe	25	*Sow*	5		

Storage/Processing
A fluffy product of below average bulk density.

Alternative Names

Bulk Density
600 kg/m³

Typical Analysis

Dry Matter	90.0	NCDG	68.0	DUP (@ 5)	3.8	Avail Lysine	-
Crude Protein	24.5	NDF	51.2	DUP (@ 8)	4.9	Methionine	0.42
DCP	20.0	ADF	11.6	Salt	0.3	Meth & Cysteine	0.7
MER	11.2	Starch	16.1	Ca	0.25	Tryptophan	0.45
MEP	9.6	Sugar	0.4	Total Phos	0.53	Threonine	0.9
DE	9.4	Starch + Sugars	16.5	Av Phos	0.18	Arginine	1.0
Crude Fibre	12.2	FME	12.2	Magnesium	0.15	PDIA	9.2
Oil (EE)	2.2	ERDP (@ 2)	18.7	Potassium	0.25	PDIN	16.3
Oil (AH)	2.5	ERDP (@ 5)	16.5	Sodium	0.25	PDIE	14.6
EFA	0.5	ERDP (@ 8)	15.3	Chloride	0.55	Met DI	0.3
Ash	5.3	DUP (@ 2)	1.6	Total Lysine	1.5	Lys DI	1.1

Starch 16.1% — NDF 51.2% — Other 0%
Sugars 0.4% — Ash 5.3%
Protein 24.5% — Oil 2.5%

Roots, Fruits and By-Products	

Introduction

Roots of Manihot esculenta crantz, regardless of their presentation.

Manioc is a tuberous root of a sub-tropical shrub which is processed before feeding to destroy the cyanide present. It is grown for its starch content and the roots are peeled, chopped and dried after harvesting. The material may come as a meal or pellet, depending on processing method. Its usage depends on price and availability of cereal. Availability may also be affected in Europe by import quotas.

Origin/Manufacture

Tropical and sub-tropical Far East.

Nutritional Benefit

Low in protein and oil but high in starch. The protein is heavily made up of non-protein nitrogen (up to 35%). The analysis will also vary depending on the extent of processing. Ideal for ruminants as starch is slowly degraded.

Colour/Texture

Muddy white meal/pellet or chips.

Palatability

Can vary depending on cyanide content.

Limits to Usage (Anti-Nutritional Factors)

Linamarin (a glucoside) present releases cyanamide. Hydrocyanic acid is limited by law and users should consider permitted levels in the Feeding Stuff Regulations.

Concentrate Inclusion % per species

Inc %		Inc %		Inc %	
Calf	5	Creep	0	Chick	5
Dairy	30	Weaner	10	Broiler	10
Beef	30	Grower	15	Breeder	10
Lamb	5	Finisher	30	Layer	15
Ewe	30	Sow	25		

Storage/Processing

Alternative Names

Cassava, Tapioca, Manihot.

Bulk Density

Typical Analysis

Dry Matter	87.0	NCGD	80.1	DUP (@ 5)	0.45	Avail Lysine	0.05
Crude Protein	3.0	NDF	15.4	DUP (@ 8)	0.63	Methionine	0.05
DCP	1.1	ADF	6.4	Salt	0.2	Meth & Cysteine	0.07
MER	13.2	Starch	71.0	Ca	0.2	Tryptophan	0.03
MEP	14.9	Sugar	3.0	Total Phos	0.2	Threonine	0.07
DE	15.15	Starch + Sugars	74.0	Av Phos	0.15	Arginine	0.15
Crude Fibre	5.0	FME	13.3	Magnesium	0.15	PDIA	0.8
Oil (EE)	0.6	ERDP (@ 2)	2.1	Potassium	1.1	PDIN	1.9
Oil (AH)	1.4	ERDP (@ 5)	1.8	Sodium	0.05	PDIE	8.5
EFA	-	ERDP (@ 8)	1.6	Chloride	0.15	Met DI	0.02
Ash	6.2	DUP (@ 2)	0.16	Total Lysine	0.1	Lys DI	0.05

- Starch 71%
- Sugars 3%
- Protein 3%
- NDF 15.4%
- Ash 6.2%
- Oil 1.4%
- Other 0%

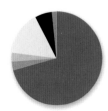

Miscellaneous	

Introduction

Products obtained by heating, drying and grinding whole or parts of warm blooded land animals from which the fat has been partially extracted or physically removed. The product must be substantially free of hooves, horn, bristle, hair and feathers as well as digestive tract content.

A by-product from the rendering industry made from scraps of meat, trimmings, bones and condensed carcasses. They are chopped and cooked, the oil is extracted and they are then ground and screened.

Origin/Place of Manufacture

Produced throughout the world, but usually consumed in country of production.

Nutritional Benefit.

High protein (+45%) and high ash from the bone, resulting in good levels of calcium, phosphorus and magnesium. Analysis varies between production plants, animals processed, etc. If overheated, protein availability will be reduced.

Colour/Texture

Brown textured gritty meal.

Palatability

Average for pigs and poultry.

Limits to Usage (Anti-Nutritional Factors)

Legislation and consumer demands means meat and bone meal is not used within the UK and many other countries around the world. MAFF legislation must be adhered to.

Concentrate Inclusion % per species

	Inc %		Inc %		Inc %
Calf	0	Creep	0	Chick	2.5
Dairy	2.5	Weaner	0	Broiler	2.5
Beef	2.5	Grower	2.5	Breeder	2.5
Lamb	0	Finisher	2.5	Layer	2.5
Ewe	2.5	Sow	2.5		

Storage/Processing

Depending on oil content, the product may go rancid very quickly.

Alternative Names

Bulk Density

625 - 725 Kg/m³

Typical Analysis

Dry Matter	91.0	NCDG	55.0	DUP (@ 5)	-	Avail Lysine	-
Crude Protein	54.0	NDF	2.5	DUP (@ 8)	-	Methionine	1.2
DCP	43.0	ADF	-	Salt	1.4	Meth & Cysteine	0.5
MER	12.4	Starch	0	Ca	9.1	Tryptophan	0.5
MEP	11.6	Sugar	0	Total Phos	4.7	Threonine	1.5
DE	12.9	Starch + Sugars	0	Av Phos	4.2	Arginine	3.0
Crude Fibre	2.5	FME	7.8	Magnesium	0.3	PDIA	22.0
Oil (EE)	13.0	ERDP (@ 2)	-	Potassium	0.6	PDIN	34.0
Oil (AH)	13.5	ERDP (@ 5)	-	Sodium	0.8	PDIE	29.3
EFA	0.6	ERDP (@ 8)	-	Chloride	0.75	Met DI	-
Ash	30	DUP (@ 2)	-	Total Lysine	2.5	Lys DI	-

- Starch 0%
- NDF 2.5%
- Other 0%
- Sugars 0%
- Ash 30%
- Protein 54%
- Oil 13.5%

Miscellaneous	

Introduction

Products obtained by drying milk.

Dried powdered milk is produced either by spray or roller drying the residue after the cream has been separated by centrifugal force.

Origin/Place of Manufacture

Produced throughout the world.

Nutritional Benefit

Excellent quality protein of high availability with good energy levels. The fat content is low (1%). Its main use as a protein source is as a supplement in the diets of simple-stomached animals, especially young pigs and poultry on a cereal diet, and is rarely used for ruminant diets. Fed to young calves pre-weaning. It is low in magnesium, iron and vitamins D, E and A, which are all normally supplemented in commercial products.

Colour/Texture

White powder.

Palatability

Good.

Limits to Usage (Anti-Nutritional Factors)

Concentrate Inclusion % per species

	Inc %		Inc %		Inc %
Calf	10	Creep	10	Chick	0
Dairy	0	Weaner	0	Broiler	0
Beef	0	Grower	0	Breeder	0
Lamb	10	Finisher	0	Layer	0
Ewe	0	Sow	0		

Storage/Processing

Dry storage essential.

Alternative Names

Bulk Density

390 - 410 Kg/m³

Milk Powder

Typical Analysis

Dry Matter	93.0	NCGD	96.0	DUP (@ 5)	2.2	Avail Lysine	2.4
Crude Protein	37.0	NDF	1.0	DUP (@ 8)	3.2	Methionine	0.8
DCP	33.0	ADF	0	Salt	1.5	Meth & Cysteine	1.3
MER	13.5	Starch	1.0	Ca	1.0	Tryptophan	0.4
MEP	11.0	Sugar	46.0	Total Phos	1.3	Threonine	1.2
DE	16.5	Starch + Sugars	47.0	Av Phos	1.2	Arginine	1.0
Crude Fibre	1.0	FME	13.2	Magnesium	0.15	PDIA	0
Oil (EE)	1.0	ERDP (@ 2)	29.5	Potassium	1.5	PDIN	19.5
Oil (AH)	1.2	ERDP (@ 5)	28.2	Sodium	0.5	PDIE	7.0
EFA	0.05	ERDP (@ 8)	27.0	Chloride	1.0	Met DI	0.15
Ash	8.5	DUP (@ 2)	1.0	Total Lysine	2.5	Lys DI	0.55

Whey Powder

Typical Analysis

Dry Matter	95.0	NCGD	92.0	DUP (@ 5)	0	Avail Lysine	2.0
Crude Protein	15.0	NDF	0	DUP (@ 8)	0	Methionine	0.4
DCP	12.0	ADF	0	Salt	2.5	Meth & Cysteine	1.0
MER	13.3	Starch	0	Ca	3.5	Tryptophan	0.5
MEP	9.5	Sugar	79.0	Total Phos	1.0	Threonine	0.8
DE	15.5	Starch + Sugars	79.0	Av Phos	1.0	Arginine	0.1
Crude Fibre	0	FME	13.6	Magnesium	0.6	PDIA	0.17
Oil (EE)	2.0	ERDP (@ 2)	11.5	Potassium	0	PDIN	3.7
Oil (AH)	1.5	ERDP (@ 5)	11.5	Sodium	2.5	PDIE	2.9
EFA	0.01	ERDP (@ 8)	11.6	Chloride	1.5	Met DI	0.65
Ash	9.3	DUP (@ 2)	0	Total Lysine	2.2	Lys DI	2.2

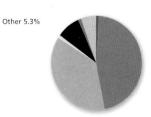

Starch 1%	NDF 1%	Other 5.3%
Sugars 46%	Ash 8.5%	
Protein 37%	Oil 1.2%	

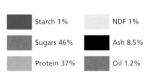

CONTEXT

Cereals and By-Products

Introduction

Grains of Panicum miliaceum L.

A grass which produces a small hard seed used for animal feed.

Origin/Place of Manufacture

Tropical and sub-tropical countries

Nutritional Benefit

High in energy, but needs to be processed prior to feeding by grinding or flaking. High in poorly digested fibre which means it is not suitable for young animals. Ideal for pigs as it produces a hard white fat.

Colour/Texture

Yellow/orange

Palatability

Very palatable

Limits to Usage (Anti-Nutritional Factors)

Better results are achieved when it is fed with other cereals. The crude fibre content is poorly digested by young stock.

Concentrate Inclusion % per species

	Inc %		Inc %		Inc %
Calf	35	Creep	20	Chick	25
Dairy	40	Weaner	25	Broiler	30
Beef	45	Grower	30	Breeder	35
Lamb	30	Finisher	40	Layer	40
Ewe	45	Sow	40		

Storage/Processing

Alternative Names

Bulk Density

600 - 675 Kg/m³

Typical Analysis

Dry Matter	89.0	NCGD	-	DUP (@ 5)	-	Avail Lysine	0.2
Crude Protein	11.0	NDF	-	DUP (@ 8)	-	Methionine	0.3
DCP	8.0	ADF	-	Salt	-	Meth & Cysteine	0.4
MER	12.3	Starch	-	Ca	0.05	Tryptophan	0.2
MEP	13.5	Sugar	-	Total Phos	0.3	Threonine	0.3
DE	14.9	Starch + Sugars	-	Av Phos	0.1	Arginine	0.4
Fibre	6.5	FME	-	Magnesium	0.15	PDIA	-
Oil (EE)	3.5	ERDP (@ 2)	-	Potassium	0.4	PDIN	-
Oil (AH)	3.5	ERDP (@ 5)	-	Sodium	0.05	PDIE	-
EFA	-	ERDP (@ 8)	-	Chloride	0.15	Met DI	-
Ash	3.5	DUP (@ 2)	-	Total Lysine	0.25	Lys DI	-

Starch -   NDF -   Other 75.5%

Sugars -   Ash 3.5%

Protein 11%   Oil 3.5%

Miscellaneous

Introduction

The 'family' name given to a wide variety of minerals added to balance raw materials in finished feeds. Most are made of inorganic forms. Supplementary minerals provide the deficit between the animal requirements and that provided by the raw materials. Mineral supplements are either tailor made to suit specific species, age and production level eg pig and poultry supplemented or formulated to a range which suits most diets eg. Ruminant High Magnesium, Dry Cow and General Purpose supplements.

Origin/Place of Manufacture

Throughout Europe and the rest of the world.

Nutritional Benefit

Essential for optimal performance of todays highly productive stock and preventing a range of possible deficiency syndromes. Usually formulated to supply major and minor minerals plus vitamins.

Colour/Texture

From grey to brown.

Palatability

Average. Salt included in free access product to encourage intake.

Limits to Usage (Anti-Nutritional Factors)

In feed minerals cannot easily be fed free access due to their dustiness and the fact that they get washed away with rain. No additional copper mineral essential for sheep.

Concentrate Inclusion % per species as recommended by supplier

	Inc %		Inc %		Inc %
Calf	0.25	Creep	0.25	Chick	0.25
Dairy	0.25	Weaner	0.25	Broiler	0.25
Beef	0.25	Grower	0.25	Breeder	0.25
Lamb	0.25	Finisher	0.25	Layer	0.25
Ewe	0.25	Sow	0.25		

Storage/Processing

Store in a cool dry building. Use within 6 months if possible.

Alternative Names

Bulk Density

800 kg/m³

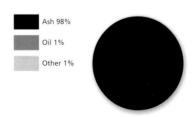

- Ash 98%
- Oil 1%
- Other 1%

Roots, Fruits and By-Products

Introduction

By-Product consisting of the syrupy residue collected during the manufacture of refining of beet sugar.

A by-product from sugar manufactured from sugar beet. The beet are crushed and the sugars removed into warm water. After crystallisation from the water extract, a thick black liquid (molasses) remains. It is usually added back to the beet pulp with quantities sold separately as a liquid.

Origin/Manufacture

Northern Europe, USA.

Nutritional Benefit

This product contains 70-75% DM of which 50% consists of sugars. The molasses has a crude protein content of 4-7% with most of this being in the form of non-protein nitrogen compound. These include amines and betaines, which are responsible for the fishy smell associated with the extraction process.

Colour/Texture

Brown, viscous liquid.

Palatability

Good, but not as sweet tasting as cane molasses.

Limits to Usage

High in potassium and salt. Known to be a laxative to animals. High levels of sugar in ruminant ration could lead to acidosis.

Concentrate Inclusion % per species

	Inc %		Inc %		Inc %
Calf	5	Creep	1	Chick	1
Dairy	15	Weaner	3	Broiler	3
Beef	15	Grower	3	Breeder	3
Lamb	10	Finisher	4	Layer	3
Ewe	15	Sow	5		

Storage/Processing

Stores well for up to 1 year. Will thicken in cold weather. Usually stored in elevated tank and moved by gravity. Aids in pelleting and expanding process. Molasses can be added to an absorbent which includes bran, brewers grain, malt culms, etc. The nutritive value of these products is obviously highly dependent upon the absorbent used.

Alternative Names

Bulk Density

Typical Analysis

Dry Matter	75.0	NCGD	6.6	DUP (@ 5)	0.55	Avail Lysine	0.01
Crude Protein	7.5	NDF	0	DUP (@ 8)	0.65	Methionine	0.02
DCP	4.5	ADF	0	Salt	0.9	Meth & Cysteine	0.03
MER	12.2	Starch	0	Ca	0.6	Tryptophan	0.01
MEP	11.3	Sugar	66.0	Total Phos	0.1	Threonine	0.03
DE	13.9	Starch + Sugars	66.0	Av Phos	0.05	Arginine	0.01
Crude Fibre	0	FME	12.0	Magnesium	0.05	PDIA	0
Oil (EE)	0.4	ERDP (@ 2)	4.3	Potassium	4.5	PDIN	2.6
Oil (AH)	0.4	ERDP (@ 5)	3.8	Sodium	0.8	PDIE	6.5
EFA	0.2	ERDP (@ 8)	3.5	Chloride	2.0	Met DI	0.01
Ash	10.0	DUP (@ 2)	0.25	Total Lysine	0.02	Lys DI	0.01

- Starch 0%
- Sugars 66%
- Protein 7.5%
- NDF 0%
- Ash 10%
- Oil 0.4%
- Other 16.1%

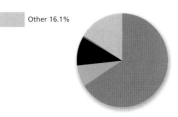

CONTEXT

Miscellaneous

Introduction

By-product consisting of the syrupy residue collected during the manufacture or refining of sugar from sugar-cane Saccharum officianarum L.

The cane is harvested, crushed and the juice removed by mechanical methods and warm water. The molasses is the residual syrup which remains when it is no longer economically viable to expend energy to crystallise out the sugars from the juices. Molasses is often mixed with condensed molasses soluble to make a free flowing higher protein blend.

Origin/Place of Manufacture

S.E. Asia, Pakistan, India, S. America, Cuba, Florida, Africa, Australia.
Blended with other liquids in port-side terminals.

Nutritional Benefit

The product is 50% sugars, making it extremely palatable to all stock. Although it is a liquid, it is very high in solids (75%). Ideal for ruminants as it is the simplest form of energy available to stimulate the rumen. It reduces dust in a mixture, aids pelleting and is a good source of minerals.

Colour/Texture

Dark brown/black viscous liquid.

Palatability

Excellent (Sweet smelling and tasting).

Limits to Usage (Anti-Nutritional Factors)

It has a high potassium and salt content which can lead to 'scouring' especially in younger animals. High levels of sugar in ruminant ration could lead to acidosis.

Concentrate Inclusion % per species

	Inc %		Inc %		Inc %
Calf	5	Creep	1	Chick	1
Dairy	15	Weaner	3	Broiler	3
Beef	15	Grower	3	Breeder	3
Lamb	10	Finisher	4	Layer	3
Ewe	15	Sow	5		

Storage/Processing

Stores well for up to 1 year. Will thicken in cold weather. Usually stored in elevated tank and moved by gravity. Aids in pelleting and expanding process. Molasses can be added to an absorbent which includes bran, brewers grain, malt culms, etc. The nutritive value of these products is obviously highly dependent upon the absorbent used.

Alternative Names

Blackstrap molasses.

Bulk Density

1400 Kg/m³

Typical Analysis

Dry Matter	75.0	NCGD	67.0	DUP (@ 5)	0.55	Avail Lysine	0.01
Crude Protein	6.0	NDF	0	DUP (@ 8)	0.65	Methionine	0.02
DCP	4.0	ADF	0	Salt	1.3	Meth & Cysteine	0.03
MER	12.7	Starch	0	Ca	0.9	Tryptophan	0.01
MEP	11.5	Sugar	65.0	Total Phos	0.15	Threonine	0.05
DE	14.1	Starch + Sugars	65.0	Av Phos	0.07	Arginine	0.01
Crude Fibre	0	FME	12.2	Magnesium	0.5	PDIA	0
Oil (EE)	0.2	ERDP (@ 2)	3.9	Potassium	4.0	PDIN	2.6
Oil (AH)	0.2	ERDP (@ 5)	3.6	Sodium	0.3	PDIE	6.5
EFA	0	ERDP (@ 8)	3.3	Chloride	2.0	Met DI	0.01
Ash	9.0	DUP (@ 2)	0.25	Total Lysine	0.02	Lys DI	0.02

Starch 0% · NDF 0% · Other 19.8% · Sugars 65% · Ash 9% · Protein 6% · Oil 0.2%

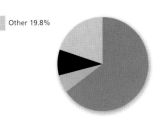

80/20 Blend (CMS - Molasses)

Typical Analysis

Dry Matter	72.0	NCGD	80.0	DUP (@ 5)	1.2	Avail Lysine	0.01
Crude Protein	10.0	NDF	0	DUP (@ 8)	1.35	Methionine	0.02
DCP	4.0	ADF	0	Salt	1.25	Meth & Cysteine	0.03
MER	12.2	Starch	0	Ca	1.0	Tryptophan	0.01
MEP	9.0	Sugar	53.0	Total Phos	0.15	Threonine	0.05
DE	11.6	Starch + Sugars	53.0	Av Phos	0.7	Arginine	0.01
Crude Fibre	0	FME	10.8	Magnesium	0.6	PDIA	0.1
Oil (EE)	0.1	ERDP (@ 2)	7.2	Potassium	5.0	PDIN	4.8
Oil (AH)	0.1	ERDP (@ 5)	6.0	Sodium	0.5	PDIE	6.6
EFA	0	ERDP (@ 8)	6.0	Chloride	2.5	Met DI	0.01
Ash	15.0	DUP (@ 2)	0.7	Total Lysine	0.02	Lys DI	0.02

Condensed Molasses Solubles (CMS)

Typical Analysis

Dry Matter	60.0	NCGD	-	DUP (@ 5)	-	Avail Lysine	-
Crude Protein	35.0	NDF	-	DUP (@ 8)	-	Methionine	-
DCP	-	ADF	-	Salt	4.8	Meth & Cysteine	-
MER	10.0	Starch	0	Ca	1.5	Tryptophan	-
MEP	-	Sugar	5.0	Total Phos	0.65	Threonine	-
DE	8.3	Starch + Sugars	-	Av Phos	0.25	Arginine	-
Crude Fibre	-	FME	-	Magnesium	0.2	PDIA	-
Oil (EE)	0.5	ERDP (@ 2)	-	Potassium	7.0	PDIN	-
Oil (AH)	0.5	ERDP (@ 5)	-	Sodium	2.5	PDIE	-
EFA	-	ERDP (@ 8)	-	Chloride	3.5	Met DI	-
Ash	23.3	DUP (@ 2)	-	Total Lysine	-	Lys DI	-

Oil Seeds and By-Products

Introduction

By-product of oil manufacture, obtained by pressing of seeds of the niger plant Guizotia abyssinica (Lf) Cass.

The small and black seeds of the herbaceous Niger plant are grown for their oil, which is extracted by solvent or by pressing. It is rarely available commercially.

Origin/Place of Manufacture

Africa.

Nutritional Benefit

High in fibre and crude protein but of low digestibility, with limited nutritive value. Also low in Lysine.

Colour/Texture

Palatability

Poor.

Limits to Usage (Anti-Nutritional Factors)

The low digestibility and poor palatability limits its inclusion in livestock rations.

Concentrate Inclusion % per species

	Inc %		Inc %		Inc %
Calf	0	Creep	0	Chick	0
Dairy	2.5	Weaner	0	Broiler	0
Beef	2.5	Grower	0	Breeder	0
Lamb	0	Finisher	0	Layer	2
Ewe	2.5	Sow	2.5		

Storage/Processing

Alternative Names

Bulk Density

Typical Analysis

Dry Matter	90.0	NCDG	-	DUP (@ 5)	-	Avail Lysine	-
Crude Protein	35.0	NDF	40.9	DUP (@ 8)	-	Methionine	0.6
DCP	24.0	ADF	-	Salt	0.5	Meth & Cysteine	1.3
MER	10.8	Starch	2.0	Ca	0.1	Tryptophan	0.45
MEP	13.2	Sugar	4.0	Total Phos	0.5	Threonine	1.1
DE	-	Starch + Sugars	6.0	Av Phos	0.4	Arginine	-
Crude Fibre	18.5	FME	-	Magnesium	0.5	PDIA	-
Oil (EE)	8.1	ERDP (@ 2)	-	Potassium	1.2	PDIN	-
Oil (AH)	-	ERDP (@ 5)	-	Sodium	0.05	PDIE	-
EFA	-	ERDP (@ 8)	-	Chloride	0.25	Met DI	-
Ash	10.0	DUP (@ 2)	-	Total Lysine	1.1	Lys DI	-

- Starch 2%
- NDF 40.9%
- Other 0%
- Sugars 4%
- Ash 10%
- Protein 35%
- Oil 8.1%

Cereals and By-Products	

Introduction

By-Product obtained during the processing of screened, dehusked oats into oat groats and flour. It consists principally of oat bran and some endosperm.

Oats are widely used in porridge and other breakfast products. Oat feed is produced as a by-product of processing porridge and consists of a mixture of hulls and meal remaining from the screening and dehulling process.

Origin/Place of Manufacture

UK.

Nutritional Benefit

A feed of low nutritive value but suited to ruminant feed. It has a high level of neutral detergent fibre (NDF) and a variable composition which is dependant on the levels of included hulls, flour and screen dust. Oat feed needs careful mineral supplementation.

Colour/Texture

Silvery yellow fibrous feed.

Palatability

Palatable.

Limits to Usage (Anti-Nutritional Factors)

Low energy value and of generally low nutritive value.

Concentrate Inclusion % per species

	Inc %		Inc %		Inc %
Calf	0	Creep	0	Chick	0
Dairy	15	Weaner	0	Broiler	0
Beef	15	Grower	2.5	Breeder	0
Lamb	10	Finisher	2.5	Layer	5
Ewe	10	Sow	5		

Storage/Processing

Alternative Names

Oat Middlings.

Bulk Density

400 - 430 Kg/m³

Typical Analysis

Dry Matter	89.0	NCGD	35.0	DUP (@ 5)	0.75	Avail Lysine	0.15
Crude Protein	5.0	NDF	79.0	DUP (@ 8)	0.82	Methionine	0.05
DCP	2.5	ADF	41.0	Salt	0.2	Meth & Cysteine	0.1
MER	6.0	Starch	7.5	Ca	0.1	Tryptophan	0.05
MEP	2.5	Sugar	2.0	Total Phos	0.25	Threonine	0.15
DE	5.0	Starch + Sugars	9.5	Av Phos	0.1	Arginine	0.25
Crude Fibre	29.5	FME	6.0	Magnesium	0.1	PDIA	1.0
Oil (EE)	2.0	ERDP (@ 2)	3.0	Potassium	0.5	PDIN	2.9
Oil (AH)	2.0	ERDP (@ 5)	2.5	Sodium	0.05	PDIE	7.0
EFA	0.6	ERDP (@ 8)	2.4	Chloride	0.15	Met DI	0.05
Ash	4.5	DUP (@ 2)	0.6	Total Lysine	0.2	Lys DI	0.15

Starch 7.5%	NDF 79%	Other 0%
Sugars 2%	Ash 4.5%	
Protein 5%	Oil 2%	

Cereals and By-Products

Introduction

By-product obtained during the processing of screened oats into oat groats. It consists principally of oat bran and some endosperm.

This is the husk removed from the angular oat grain, comprising 25% of its weight while oat flour meal is a mixture of both the remaining oat kernel and fragments from processing.

Origin/Place of Manufacture

UK.

Nutritional Benefit

The hulls are very fluffy, high in fibre, and low in energy. This means they are unsuitable for pigs and poultry.

Crude protein levels are extremely low (2%) and the protein is of limited digestibility.

Colour/Texture

Grey/white meal.

Palatability

Average.

Limits to Usage (Anti-Nutritional Factors)

Products low bulk density can limit inclusion.

Concentrate Inclusion % per species

	Inc %		Inc %		Inc %
Calf	5	Creep	0	Chick	0
Dairy	15	Weaner	0	Broiler	0
Beef	25	Grower	2.5	Breeder	0
Lamb	5	Finisher	2.5	Layer	2.5
Ewe	15	Sow	5		

Storage/Processing

Does not pellet well.

Alternative Names

Bulk Density

Hulls 100 - 150 Kg/m³   **Flour Meal** 320 - 352 Kg/m³

Typical Analysis

Dry Matter	89.0	NCGD	24.0	DUP (@ 5)	-	Avail Lysine	-
Crude Protein	6.0	NDF	43.0	DUP (@ 8)	-	Methionine	-
DCP	5.0	ADF	-	Salt	0.2	Meth & Cysteine	-
MER	4.5	Starch	10.0	Ca	0.2	Tryptophan	-
MEP	2.0	Sugar	6.0	Total Phos	0.3	Threonine	-
DE	3.0	Starch + Sugars	16.0	Av Phos	-	Arginine	-
Fibre	2.8	FME	-	Magnesium	0.1	PDIA	-
Oil (EE)	2.0	ERDP (@ 2)	-	Potassium	0.1	PDIN	-
Oil (AH)	3.0	ERDP (@ 5)	-	Sodium	0.05	PDIE	-
EFA	-	ERDP (@ 8)	-	Chloride	0.16	Met DI	-
Ash	4.0	DUP (@ 2)	-	Total Lysine	-	Lys DI	-

Starch 10%   NDF 43%   Other 0%

Sugars 6%   Ash 4%

Protein 6%   Oil 3%

Cereals and By-Products	

Introduction

Grains of Avena sativa L. and other cultivars of oats.

A husky seed grown largely for use in foods such as breakfast cereals. The fibrous husk can weigh up to 25% of the seed weight and is sometimes removed before feeding. The husk content varies with variety and season with spring varieties higher in fibre. They are usually rolled or flaked to enhance the digestibility.

Origin/Place of Manufacture

Grown in temperate countries, especially the UK.

Nutritional Benefit

Poorer in energy value than wheat and barley, but higher in unsaturated oil. Popular in ruminant diets compared to non-ruminant due to the high fibre content. Lysine, methionine and histidine are low. Can produce good quality carcass fat and reduce milk butter fat due to unsaturated oil level. Less suitable for young stock than the other cereals, with growth reduction having been observed.

As they are fibrous, they are an ideal cereal for horses. The oil content is high compared to other cereal (64%) and it is rich in polyunsaturated fatty acids (PUFAs). The rolled form is often included in coarse feeds.

Colour/Texture

Cream coloured clothed seed.

Palatability

Very palatable.

Limits to Usage (Anti-Nutritional Factors)

High in phytic acid which encourages phosphorus binding. Protein is also high in glutamic acid. The oil content is unsaturated and could, therefore, result in soft carcass fat.

Concentrate Inclusion % per species

	Inc %		Inc %		Inc %
Calf	10	Creep	0	Chick	0
Dairy	25	Weaner	15	Broiler	0
Beef	35	Grower	15	Breeder	15
Lamb	10	Finisher	25	Layer	15
Ewe	25	Sow	25		

Storage/Processing

Alternative Names

Naked Oats.

Bulk Density

450 - 550 Kg/m³   1000 grain weight = 35 - 40 grams

Typical Analysis

Dry Matter	86.0	NCDG	65.0	DUP (@ 5)	0.6	Avail Lysine	0.35
Crude Protein	11.8	NDF	37.6	DUP (@ 8)	0.8	Methionine	0.23
DCP	9.4	ADF	12.8	Salt	0.17	Meth & Cysteine	0.6
MER	12.2	Starch	43.0	Ca	0.12	Tryptophan	0.1
MEP	12.5	Sugar	1.1	Total Phos	0.35	Threonine	0.45
DE	13.2	Starch + Sugars	44.1	Av Phos	0.12	Arginine	0.8
Crude Fibre	10.0	FME	10.0	Magnesium	0.17	PDIA	2.9
Oil (EE)	4.0	ERDP (@ 2)	10.3	Potassium	0.47	PDIN	8.2
Oil (AH)	4.5	ERDP (@ 5)	9.9	Sodium	0.06	PDIE	8.4
EFA	2.1	ERDP (@ 8)	9.7	Chloride	0.12	Met DI	0.1
Ash	2.0	DUP (@ 2)	0.4	Total Lysine	0.52	Lys DI	0.5

Starch 43%   NDF 37.6%   Other 0%

Sugars 1.1%   Ash 2%

Protein 11.8%   Oil 4.5%

CONTEXT

Oilseeds and By-Products

Introduction

By-product of oil manufacture, obtained by extraction of pressed olives (Olea europaea L.) separated as far as possible from parts of the kernel.

Olives are processed for the oil which is seen as a 'healthy' vegetable oil. Olive pulp is the residue from oil extraction.

Origin/Place of Manufacture

Grown in the Mediterranean, Northern African and parts of America.

Nutritional Benefit

A digestible fibre feed of variable nutritional quality due to different processing techniques. Unsuitable for pigs and poultry due to high fibre content. Low to medium protein content and deficient in minerals.

Colour/Texture

Dark brown pellets.

Palatability

Low.

Limits to Usage (Anti-Nutritional Factors)

Palatability will limit inclusion rates.

Concentrate Inclusion % per species

	Inc %		Inc %		Inc %
Calf	0	Creep	0	Chick	0
Dairy	10	Weaner	0	Broiler	0
Beef	10	Grower	0	Breeder	0
Lamb	0	Finisher	0	Layer	0
Ewe	10	Sow	0		

Storage/Processing

Alternative Names

Bulk Density

700 - 750 Kg/m³

Typical Analysis

Dry Matter	88.0	NDF	59.2	DUP (@ 8)	1.9	Methionine	-
Crude Protein	11.0	ADF	50.0	Salt	0.7	Meth & Cysteine	-
DCP	8.0	Starch	0.3	Ca	1.5	Tryptophan	-
MER	5.1	Sugar	0.3	Total Phos	0.1	Threonine	-
MEP	-	Starch + Sugars	0.6	Av Phos	-	Arginine	-
DE		FME	3.5	Magnesium	0.1	PDIA	-
Crude Fibre	20.0	ERDP (@ 2)	6.9	Potassium	0.5	PDIN	-
Oil (EE)	1.5	ERDP (@ 5)	6.6	Sodium	3.0	PDIE	-
Oil (AH)	1.7	ERDP (@ 8)	6.2	Chloride	-	Met DI	-
Ash	7.5	DUP (@ 2)	1.1	Total Lysine	-	Lys DI	-
NCGD	34.0	DUP (@ 5)	1.6	Avail Lysine	-		

Starch 0.3%	NDF 59.2%	Other 0%
Sugars 0.3%	Ash 7.5%	
Protein 11%	Oil 1.7%	

Oilseeds and By-Products

Introduction

By-product of oil manufacture, obtained by expelling and/or extraction of palm kernels from which as much as possible of the hard shell has been removed.

Large fruit berries are harvested from the palm tree then crushed, expelled or extracted to produce palm oil and palm kernel meal. Palm oil is used in foods and soaps.

Origin/Place of Manufacture

Asia, West Africa.

Nutritional Benefit

A source of protein and energy with a high fibre level, most widely used in ruminant diets. It will contain some palm oil which is a hard oil, producing hard carcass fat. Not suitable for use on its own but often mixed with molasses to encourage intake. Low in lysine and other amino acids, except methionine. Analysis can be variable if source unknown. High inclusion levels can help to boost butter fat levels.

Colour/Texture

Pale brown, dry gritty meal with a soapy smell.

Palatability

Poor.

Limits to Usage (Anti-Nutritional Factors)

May contain high levels of aflatoxin and salmonella so seek supplier assurances. Some sources have small fragments remaining that can puncture cattle hooves.

Concentrate Inclusion % per species

	Inc %		Inc %		Inc %
Calf	5	Creep	0	Chick	0
Dairy	20	Weaner	0	Broiler	0
Beef	20	Grower	2.5	Breeder	0
Lamb	5	Finisher	5.0	Layer	0
Ewe	10	Sow	2.5		

Storage/Processing

Can only be stored for a limited period.

Alternative Names

Bulk Density

380 - 450 Kg/m³

Typical Analysis	Expeller	Extraction	Typical Analysis	Expeller	Extraction
Dry Matter	89.0	89.0	DUP (@ 5)	6.7	6.3
Crude Protein	18.0	18.0	DUP (@ 8)	7.9	7.7
DCP	14.2	14.0	Salt	0.25	0.2
MER	12.8	12.3	Ca	0.25	0.25
MEP	8.1	7.9	Total Phos	0.65	0.65
DE	12.9	12.5	Av Phos	0.2	0.2
Crude Fibre	17.5	16.5	Magnesium	0.3	0.3
Oil (EE)	8.3	2.0	Potassium	0.6	0.6
Oil (AH)	9.2	2.3	Sodium	0.02	0.02
EFA	0.25	0.1	Chloride	0.5	0.5
Ash	4.5	4.5	Total Lysine	0.6	0.6
NCGD	66.0	63.0	Avail Lysine	0.5	0.5
NDF	68.0	67.1	Methionine	0.4	0.4
ADF	46.0	42.1	Meth & Cysteine	0.7	0.7
Starch	4.0	4.1	Tryptophan	0.2	0.2
Sugar	1.75	3.0	Threonine	0.6	0.6
Starch + Sugars	5.75	7.2	Arginine	1.0	1.0
FME	9.8	10.9	PDIA	11.5	11.5
ERDP (@ 2)	11.6	11.2	PDIN	14.3	14.3
ERDP (@ 5)	8.5	8.2	PDIE	16.5	16.5
ERDP (@ 8)	7.5	7.2	Met DI	0.3	0.3
DUP (@ 2)	3.5	3.3	Lys DI	0.6	0.6

Starch 4.1% NDF 68.1% Other 0%

Sugars 3% Ash 5%

Protein 18% Oil 1.8%

Forages and Stock Feeds	

Introduction

The parsnip is a root vegetable and, when surplus or rejected parsnips are available, they are sold for animal feed.

Origin/Manufacture

UK.

Nutritional Benefit

Ideal when fresh and will encourage forage intake. They are slightly higher in Dry Matter (15%) than carrots and slightly higher in energy (13 MJ/Kg DM). Starch and sugar levels are approximately 11%.

Colour/Texture

White/grey, whole/broken roots.

Palatability

Good when fresh.

Limits to Usage

Forage Inclusion % per species

	Inc %		Inc %		Inc %
Calf	0	Creep	0	Chick	0
Dairy	15.0	Weaner	0	Broiler	0
Beef	15.0	Grower	0	Breeder	0
Lamb	0	Finisher	0	Layer	0
Ewe	5.0	Sow	0		

Storage/Processing

Feed within 2 weeks.

Alternative Names

Bulk Density

Typical Analysis

Dry Matter	15.0	NDF	47.5	DUP (@ 8)	2.9	Tryptophan	-
Crude Protein	8.5	ADF	13.0	Salt	0.7	Threonine	-
DCP	6.5	Starch	3.0	Ca	0.05	Arginine	-
MER	13.0	Sugar	8.0	Total Phos	0.3	PDIA	-
MEP	-	Starch + Sugars	11.0	Magnesium	0.1	PDIN	-
Crude Fibre	7.5	FME	12.3	Potassium	1.0	PDIE	-
Oil (EE)	1.5	ERDP (@ 2)	6.5	Sodium	0.05	Met DI	-
Oil (AH)	1.8	ERDP (@ 5)	5.5	Chloride	-	Lys DI	-
EFA	-	ERDP (@ 8)	4.0	Total Lysine	-		
Ash	5.0	DUP (@ 2)	1.5	Methionine	-		
NCGD	-	DUP (@ 5)	2.0	Meth & Cysteine	-		

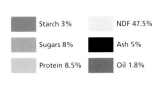

- Starch 3%
- Sugars 8%
- Protein 8.5%
- NDF 47.5%
- Ash 5%
- Oil 1.8%
- Other 26.2%

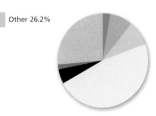

CONTEXT

Legumes and by-products	

Introduction

Seeds of Pisum spp.

Plants are harvested when young and juicy for human consumption, while those for animal feed are combine harvested and are usually harder, drier and more mature. Combined field peas are used for human consumption, for soaking and canning or as dried peas. Peas are usually steamed, flaked or micronised before use in animal feed to improve the digestibility.

Origin/Place of Manufacture

Temperate countries.

Nutritional Benefit

Often regarded as similar in analysis to field beans, but marginally lower in protein (26%). Can be used to replace soya and other protein sources but slightly lower in energy. Peas are also high in sugar and starch which makes them attractive for ruminant rations. Low in methionine, cysteine and tryptophan, but can supply some undegradable protein if heated for long periods. Oil present is unsaturated, with B vitamins present in reasonable quantities. Approximately 15% of the starch present is rumen unfermented.

Colour/Texture

Light green. Usually fed as a meal or in flakes.

Palatability

Average, but limiting to inclusion in pig and poultry diets.

Limits to Usage (Anti-Nutritional Factors)

Potential trypsin-inhibitors and/or phyto haemagglutinins (lectins, etc.) present. Tannins are sometimes found in the seed coat. Heating destroys most of the anti-nutritive factors and new varieties have reduced the levels significantly. Can have a laxative effect.

Concentrate Inclusion % per species

	Inc %		Inc %		Inc %
Calf	10	Creep	0	Chick	0
Dairy	30	Weaner	5.0	Broiler	0
Beef	30	Grower	10	Breeder	7.5
Lamb	10	Finisher	17.5	Layer	7.5
Ewe	25	Sow	17.5		

Storage/Processing

Above 25% will not pellet well.

Alternative Names

Bulk Density

725 - 800 Kg/m³

Typical Analysis

Dry Matter	86.0	NCGD	93.5	DUP (@ 5)	2.6	Avail Lysine	1.5
Crude Protein	26.0	NDF	19	DUP (@ 8)	3.8	Methionine	0.25
DCP	23.0	ADF	7.5	Salt	0.2	Meth & Cysteine	0.55
MER	13.6	Starch	43.5	Ca	0.1	Tryptophan	0.2
MEP	13.0	Sugar	6.0	Total Phos	0.6	Threonine	0.9
DE	15.4	Starch + Sugars	49.5	Av Phos	0.15	Arginine	2.4
Crude Fibre	7.0	FME	13.1	Magnesium	0.15	PDIA	2.5
Oil (EE)	1.6	ERDP (@ 2)	21.3	Potassium	1.1	PDIN	15.5
Oil (AH)	2.0	ERDP (@ 5)	20.5	Sodium	0.05	PDIE	9.5
EFA	0.7	ERDP (@ 8)	19.5	Chloride	0.05	Met DI	0.2
Ash	3.5	DUP (@ 2)	1.5	Total Lysine	1.8	Lys DI	0.7

- Starch 43.5%
- Sugars 6%
- Protein 26%
- NDF 19%
- Ash 3.5%
- Oil 2%
- Other 0%

Cereals and By-Products

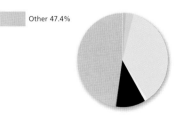

Introduction

This is the rich liquid remaining after the first distillation in malt whisky production. It contains spent yeast and unfermented soluble components. The liquid is drawn off and concentrated by evaporation. If mixed with draff and dried, the result is distillers dark grains. If not added back to the grains, it is often used as a straight liquid feed or mixed with high sugar products, eg. molasses.

Origin/Place of Manufacture

Scotland.

Nutritional Benefit

A highly palatable and nutritious liquid feed for ruminants with a salty taste and malty smell which is known to encourage forage intake. Contains highly rumen degradable protein and lysine. It has been claimed to improve digestion and utilisation of low protein fibrous feeds. High in phosphorus, magnesium and other minerals.

Pot ale syrup is an ideal supplement in many beef and sheep diets, especially for feeding with straw or low protein roughage. High ME content (14.2 MJ/kg DM) and high crude protein content (35 % DM). It has been fed successfully to pigs in liquid feed systems in quantities up to 30% of their Dry Matter intake.

Colour/Texture

Golden brown viscous liquid (variable).

Palatability

Good.

Limits to Usage (Anti-Nutritional Factors)

High in copper (avoid feeding to sheep already on other copper supplements), with a low pH 3.5 - 4.0. These factors reduce inclusion rates. High in potassium which may cause scouring. Can be highly viscous.

Concentrate Inclusion % per species

	Inc %		Inc %		Inc %
Calf	10	Creep	0	Chick	0
Dairy	25	Weaner	0	Broiler	0
Beef	20	Grower	10	Breeder	0
Lamb	0	Finisher	0	Layer	0
Ewe	10	Sow	10		

Storage/Processing

Stores well as it is acidic.

Alternative Names

Burnt Ale Syrup, Barley Distillers Solubles.

Bulk Density

1050 - 1150 Kg/m³

Typical Analysis

Dry Matter	45.0	NCDG	79.0	DUP (@ 5)	9.5	Avail Lysine	1.0
Crude Protein	37.0	NDF	0.7	DUP (@ 8)	14.6	Methionine	0.3
DCP	28.0	ADF	0.3	Salt	0.2	Meth & Cysteine	0.8
MER	14.2	Starch	1.3	Ca	0.15	Tryptophan	0.3
MEP	11.5	Sugar	2.8	Total Phos	2.1	Threonine	1.1
DE	12.0	Starch + Sugars	4.1	Av Phos	1.6	Arginine	1.0
Crude Fibre	0.4	FME	13.3	Magnesium	0.6	PDIA	14.5
Oil (EE)	0.2	ERDP (@ 2)	29.0	Potassium	2.2	PDIN	25.1
Oil (AH)	0.25	ERDP (@ 5)	28.8	Sodium	0.1	PDIE	21.0
EFA	0.2	ERDP (@ 8)	28.5	Choride	0.1	Met DI	0.1
Ash	10.5	DUP (@ 2)	3.5	Total Lysine	1.5	Lys DI	0.05

Starch 1.3%	NDF 0.7%	Other 47.4%
Sugars 2.8%	Ash 10.5%	
Protein 37%	Oil 0.3%	

Roots, Fruits and By-Products

Introduction

Solanum tuberosum L.

Grown as a tuber for human food, with excess and substandard qualities sold as 'stock-feed'. When available, they make an ideal supplement to forage and can even be fed to older pigs.

Origin/Place of Manufacture

UK.

Nutritional Benefit

A good ruminant feed when available. Contain good energy and excellent starch levels (60%). The protein level is 10-11%, with half of this being in the form of non-protein nitrogen compounds. Having a low fibre level, they are particularly suitable for pigs and poultry but need to be cooked. Potatoes are a poor source of minerals except potassium. Approximately 20% of its phosphorus content is in the form of phytates.

Colour/Texture

White/whole or part tubers.

Palatability

Avoid rotten, dirty potatoes for good palatability.

Limits to Usage (Anti-Nutritional Factors)

High in starch, so over feeding may encourage acidosis. Rotten, green and dirty potatoes should be avoided as they may contain alkaloids. Small potatoes produce a risk of animals choking, and chopping helps. Sprouted potatoes are dangerous to livestock. Avoid soil contamination.

Concentrate Inclusion % per species

	Inc %		Inc %		Inc %
Calf	0	Creep	0	Chick	0
Dairy	12	Weaner	10	Broiler	0
Beef	12	Grower	15	Breeder	0
Lamb	0	Finisher	15	Layer	0
Ewe	3	Sow	15		

Storage/Processing

Cooked potatoes are better for pigs. Where possible, feed from the ground. Green potatoes and shoots from potatoes should be avoided.

Bulk Density

Typical Analysis

Dry Matter	20.0	NDF	13.3	DUP (@ 8)	1.8	Methionine	0.12
Crude Protein	11.0	ADF	0	Salt	0.05	Meth & Cysteine	0.2
DCP	8.0	Starch	62.0	Ca	0.05	Tryptophan	0.1
MER	13.5	Sugar	8.0	Total Phos	0.2	Threonine	0.2
MEP	13.0	Starch + Sugars	70.0	Av Phos	0.1	Arginine	0.05
DE	12.0	FME	13.0	Magnesium	0.1	PDIA	2.5
Crude Fibre	4.0	ERDP (@ 2)	9.0	Potassium	2.5	PDIN	6.5
Oil (EE)	0.2	ERDP (@ 5)	8.5	Sodium	0.05	PDIE	10.5
Oil (AH)	0.2	ERDP (@ 8)	8.0	Chloride	0.05	Met DI	-
Ash	5.5	DUP (@ 2)	1.1	Total Lysine	0.3	Lys DI	-
NCGD	85.0	DUP (@ 5)	1.5	Avail Lysine	0.25		

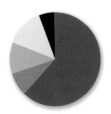

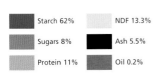

- Starch 62%
- Sugars 8%
- Protein 11%
- NDF 13.3%
- Ash 5.5%
- Oil 0.2%
- Other 0%

Potato Sludge

Roots, Fruits and By-Products	

Introduction
Produced from the processing, washing, chopping and slurrying of potatoes. When processing for chips, the removed corners and fragments are added to the sludge.

Origin/Place of Manufacture
UK.

Nutritional Benefit
High in starch but not much else. Quality depends on fibre and soil content. Ideal for ruminants and wet feeding systems for pigs. Low in Methionine.

Colour/Texture
Grey/white.

Palatability
Average.

Limits to Usage (Anti-Nutritional Factors)
Soil contamination should be avoided. The product may also be too variable to rely on.

Forage % per species

	Inc %		Inc %		Inc %
Calf	10	Creep	5	Chick	0
Dairy	25	Weaner	10	Broiler	0
Beef	30	Grower	15	Breeder	0
Lamb	10	Finisher	20	Layer	0
Ewe	25	Sow	25		

Storage/Processing
Should be fed within 7 days of delivery as unstable.

Alternative Names

Bulk Density

Typical Analysis

Dry Matter	25.0	NDF	7.0	DUP (@ 8)	1.9	Methionine	0.2
Crude Protein	7.0	ADF	0	Salt	0.05	Meth & Cysteine	0.3
DCP	5.0	Starch	78.0	Ca	0.1	Tryptophan	0.1
MER	13.3	Sugar	1.5	Total Phos	0.15	Threonine	0.4
MEP	13.0	Starch + Sugars	79.5	Av Phos	0.1	Arginine	0.4
DE	12.0	FME	13	Magnesium	0.1	PDIA	2.5
Crude Fibre	2.5	ERDP (@ 2)	6.0	Potassium	0.15	PDIN	6.5
Oil (EE)	0.8	ERDP (@ 5)	5.0	Sodium	0.02	PDIE	10.3
Oil (AH)	1.0	ERDP (@ 8)	4.5	Chloride	0.01	Met DI	-
Ash	5.5	DUP (@ 2)	10	Total Lysine	0.5	Lys DI	-
NCGD	82.0	DUP (@ 5)	14	Avail Lysine	0.3		

- Starch 78%
- Sugars 1.5%
- Protein 7%
- NDF 7%
- Ash 5.5%
- Oil 1%
- Other 0%

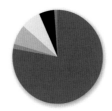

Miscellaneous	

Introduction

Products obtained by drying and grinding waste from slaughtered poultry. The product must be substantially free of feathers.

A by-product from poultry processing, consisting of hydrolysed feathers, fats, cuttings, intestines, heads, and other waste products. It is normally heat treated, dried and ground. The addition of anti-oxidants is essential to prevent oxidation.

Origin/Place of Manufacture

UK and Europe.

Nutritional Benefit

The composition varies depending on the contents added, especially that of oils and feathers. The oil level can be up to 30% of the Dry Matter, bringing the energy to 17MJ/Kg DM and ash up to 15%. The addition of feathers reduces protein availability. The product has a reasonable amino acid balance.

Colour/Texture

A grey, friable textured meal.

Palatability

Poor.

Limits to Usage (Anti-Nutritional Factors)

Legislation and consumer demand restrict it usage. MAFF legislation must be adhered to. High levels of fats can make the product unstable. If feather included, protein will be less available.

Concentrate Inclusion % per species

	Inc %		Inc %		Inc %
Calf	5	Creep	5	Chick	5
Dairy	7.5	Weaner	7.0	Broiler	5
Beef	7.5	Grower	7.5	Breeder	7.5
Lamb	5	Finisher	10	Layer	7.5
Ewe	7.5	Sow	7.5		

Storage/Processing

Alternative Name

Bulk Density

560 Kg/m³

Typical Analysis

Dry Matter	90.0	NCDG	56.0	DUP (@ 5)	18.5	Avail Lysine	2.0
Crude Protein	58.0	NDF	1.5	DUP (@ 8)	22.0	Methionine	0.7
DCP	49.0	ADF	0	Salt	0.5	Meth & Cysteine	2.8
MER	14.5	Starch	0	Ca	2.0	Tryptophan	0.5
MEP	14.5	Sugar	0	Total Phos	1.0	Threonine	2.1
DE	13.5	Starch + Sugars	0	Av Phos	0.9	Arginine	4.0
Crude Fibre	1.5	FME	7.3	Magnesium	0.1	PDIA	26.5
Oil (EE)	18.0	ERDP (@ 2)	32.0	Potassium	0.5	PDIN	42.0
Oil (AH)	20.0	ERDP (@ 5)	22.5	Sodium	0.5	PDIE	29.0
EFA	3.0	ERDP (@ 8)	18.5	Chloride	0.5	Met DI	-
Ash	10.0	DUP (@ 2)	10.0	Total Lysine	2.5	Lys DI	-

Starch 0%	NDF 1.5%	Other 10.5%
Sugars 0%	Ash 10%	
Protein 58%	Oil 20%	

Oilseeds and By-Products	

Introduction

Seeds of rape Brassica napus L. ssp. oleifera (metzg.) Sinsk., of Indian sarson Brassica napus l.var. Glauca (Roxb) O.E. Schulz and of rape Brassica ca,pestris L. ssp oleifera (Metzg) Sinsk. (Minimum botanical purity 94%).

Whole rapeseed has been historically of low interest due to the presence of erucic acid in the oil, the glucosinolates and myrosinase in the extracted meal, levels of sinapine and a resistant brown coat. However, breeding has eliminated many of these shortfalls, making it an important provider of soft oils for human food and subsequently meals for inclusion in livestock rations. It's economic viability in Europe is linked to EU incentives.

Origin/Place of Manufacture

UK, Mainland Europe, China, Canada.

Nutritional Benefit

Most of the rapeseed in livestock feeds is as the extracted meal. However, increasingly, full fat rapeseed products are being used, especially in broiler rations. Cooked full fat rapeseed is high in both protein and energy, being over 40% oil. It can be used to substitute soya bean products in growing pig and broiler diets.

Colour/Texture

Yellow black meal.

Palatability

Average.

Limits to Usage(Anti-Nutritional Factors)

Heat treatment of rapeseed deactivates many of the anti-nutritional factors and improves the nutritive value, especially for poultry.

Concentrate Inclusion % per species

	Inc %		Inc %		Inc %
Calf	0	Creep	0	Chick	0
Dairy	0	Weaner	10	Broiler	10
Beef	0	Grower	5	Breeder	5
Lamb	0	Finisher	5	Layer	5
Ewe	0	Sow	0		

Storage/Processing

High oil levels means the meal is prone to oxidation.

Alternative Names

Bulk Density

Typical Analysis

Dry Matter	90.0	NCDG	92.0	DUP (@ 5)	4.5	Avail Lysine	1.1
Crude Protein	22.0	NDF	19.7	DUP (@ 8)	5.8	Methionine	0.4
DCP	18.6	ADF	9.9	Salt	0.1	Meth & Cysteine	1.0
MER	19.1	Starch	2.5	Ca	0.4	Tryptophan	0.2
MEP	19.8	Sugar	4.8	Total Phos	0.8	Threonine	0.9
DE	19.0	Starch + Sugars	8.3	Av Phos	0.3	Arginine	1.2
Crude Fibre	7.2	FME	4.8	Magnesium	0.25	PDIA	1.6
Oil (EE)	45.0	ERDP (@ 2)	16.5	Potassium	0.8	PDIN	13.4
Oil (AH)	46.0	ERDP (@ 5)	14.3	Sodium	0.05	PDIE	2.0
EFA	13.5	ERDP (@ 8)	12.9	Chloride	0.05	Met DI	0.05
Ash	5.0	DUP (@ 2)	2.5	Total Lysine	1.5	Lys DI	0.15

Starch 2.5%   NDF 19.7%   Other 0%

Sugars 4.8%   Ash 5%

Protein 22%   Oil 46%

Oilseeds and By-Products

Introduction

By-product of oil manufacture, obtained by expelling and/or extraction of seeds of rape. (Minimum botanical purity 94%).

Rape is a Brassica grown for its 'soft oil' which is high in polyunsaturates. It is often grown as a break crop. Recently it has been grown as an industrial crop for 'environmentally friendly' fuel oils. The oil is expelled from the seed, with further solvent extraction used to lower the oil level of the final meal.

Origin/Place of Manufacture

Europe, Canada, China, India.

Nutritional Benefit

A high protein, good energy feed used to partially replace soya bean meal, although the protein is less digestible. The meal from industrial crops is usually high in glucosinolate/erucic acid.

Colour/Texture

Brown/black/yellow friable oily meal.

Palatability

Unpalatable

Limits to Usage (Anti-Nutritional Factors)

Anti-nutritional factors include erucic acid, glucosinalates, tannins and sinapine. Checks should be made on the erucic acid and glucosinolate levels as they are unpalatable and can even lead to deaths in poultry. However, these are normally low in new varieties. High levels of erucic acid and glucosinolates in rapeseed meal samples can lead to iodine deficiency and enlarged thyroid glands.

Phytic acid present can reduce mineral availability to pigs and poultry. Ruminants can easily deal with the ANFs but may find rapeseed meal slightly unpalatable.

Sinapine present can cause fishy taint in eggs but steam treatment of the meal helps to reduce this. Sinapine is thought to be associated with iron deficiency and to supply phytic acid which may reduce nutrient availability.

Concentrate Inclusion % per species

	Inc %		Inc %		Inc %
Calf	5	Creep	0	Chick	0
Dairy	25	Weaner	0	Broiler	2.5
Beef	25	Grower	2.5	Breeder	0
Lamb	5	Finisher	5	Layer	5.0
Ewe	20	Sow	2.5		

Storage/Processing

Correct processing will produce a stable product with little ANFs.

Alternative Names

Bulk Density

650 - 675 Kg/m³   **Whole Rape** 720 - 770 Kg/m³

Typical Analysis

Dry Matter	88.0	NCGD	70.0	DUP (@ 5)	6.9	Avail Lysine	1.8
Crude Protein	38.5	NDF	36.5	DUP (@ 8)	8.9	Methionine	0.8
DCP	32.0	ADF	19.1	Salt	0.07	Meth & Cysteine	1.1
MER	12.1	Starch	5.0	Ca	0.9	Tryptophan	0.4
MEP	10.5	Sugar	9.5	Total Phos	1.2	Threonine	1.6
DE	12.0	Starch + Sugars	14.5	Av Phos	0.4	Arginine	2.2
Crude Fibre	11.0	FME	10.5	Magnesium	0.5	PDIA	9.9
Oil (EE)	3.2	ERDP (@ 2)	30.1	Potassium	1.3	PDIN	24.0
Oil (AH)	3.5	ERDP (@ 5)	26.2	Sodium	0.1	PDIE	16.1
EFA	0.9	ERDP (@ 8)	23.5	Chloride	0.02	Met DI	0.3
Ash	7.0	DUP (@ 2)	4.1	Total Lysine	2.1	Lys DI	1.0

Starch 5%   NDF 36.5%   Other 0%

Sugars 9.5%   Ash 7%

Protein 38.5%   Oil 3.5%

CONTEXT

Cereals and By-Products

Introduction

By-product of the preparation of polished or glazed rice Oryza sativa L. It consists principally of undersized and/or broken grains.

Rice is the staple diet of many Asian countries. The rice has a fibrous husk and a layer of bran which is polished off. Rice bran is usually a mixture of bran germ, with limited hulls added back.

Origin/Place of Manufacture

Indian, Pakistan, S.E. Asia, S. America, with some production in the UK.

Nutritional Benefit

A variable product from different sources due to different processing methods and potential inclusion of husks. Husks should be minimised, with high levels raising the Ash value.

They are low in energy, with an oil level of 2.0% and protein levels in the order of 15%. Rice bran expellers are higher in energy, with oil levels greater than 10% but are not widely available.

Colour/Texture

Silver brown fibrous meal.

Palatability

Poor.

Limits to Usage (Anti-Nutritional Factors)

Variability in analysis may restrict usage. Oil present can encourage soft carcass fat if fed at high levels. Oil may also interfere with Vitamin E availability.

Concentrate Inclusion % per species

	Inc %		Inc %		Inc %
Calf	5	Creep	0	Chick	0
Dairy	20	Weaner	0	Broiler	0
Beef	20	Grower	2.5	Breeder	5.0
Lamb	5	Finisher	5.0	Layer	7.5
Ewe	15	Sow	10.0		

Storage/Processing

Stores well as it contains natural antioxidants, except where it contains high oil levels + 2%.

Alternative Names

Paddy Meal

Bulk Density

300 - 325 Kg/m³

Typical Analysis

Dry Matter	89.0	NCDG	52.0	DUP (@ 5)	4.2	Avail Lysine	0.5
Crude Protein	15.0	NDF	44.0	DUP (@ 8)	5.1	Methionine	0.5
DCP	7.5	ADF	25.0	Salt	0.2	Meth & Cysteine	0.6
MER	7.5	Starch	25.0	Ca	0.2	Tryptophan	0.15
MEP	6.5	Sugar	2.0	Total Phos	1.7	Threonine	0.6
DE	7.7	Starch + Sugars	27.0	Av Phos	0.3	Arginine	1.5
Crude Fibre	15.0	FME	7.1	Magnesium	0.4	PDIA	9.2
Oil (EE)	1.0	ERDP (@ 2)	10.0	Potassium	0.7	PDIN	11.6
Oil (AH)	2.0	ERDP (@ 5)	8.5	Sodium	0.05	PDIE	13.9
EFA	0.25	ERDP (@ 8)	7.5	Chloride	0.1	Met DI	0.31
Ash	12.0	DUP (@ 2)	3.0	Total Lysine	0.7	Lys DI	0.4

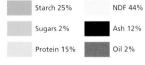

Starch 25% NDF 44% Other 0%
Sugars 2% Ash 12%
Protein 15% Oil 2%

Cereals and By-Products

Introduction

Grains of Secale cereale L.

Grown for an early crop to make seed for rye bread etc. The crop can become stemmy quickly. It can also be grown as an early forage crop.

Origin/Place of Manufacture

Asia, UK.

Nutritional Benefit

A variable product not widely used due to different processing methods from different sources and potential inclusion of husks. The inclusion of husks leads to higher Ash levels and should be minimised. It has a good energy and mineral content.

Colour/Texture

Silver brown fibrous meal.

Palatability

Poor.

Limits to Usage (Anti-Nutritional Factors)

Beware of contamination by the fungus ergot.

Concentrate Inclusion Rate % per species

	Inc %		Inc %		Inc %
Calf	10	Creep	0	Chick	2.5
Dairy	25	Weaner	0	Broiler	2.5
Beef	30	Grower	0	Breeder	2.5
Lamb	10	Finisher	0	Layer	2.5
Ewe	25	Sow	10		

Storage/Processing

Alternative Names

Bulk Density

690 Kg/m³

Typical Analysis

Dry Matter	87.0	NCGD	85.0	DUP (@ 5)	-	Avail Lysine	-
Crude Protein	11.6	NDF	38.0	DUP (@ 8)	2.6	Methionine	0.2
DCP	8.5	ADF	-	Salt	0.15	Meth & Cysteine	0.4
MER	13.5	Starch	45.0	Ca	0.05	Tryptophan	0.1
MEP	13.0	Sugar	3.0	Total Phos	0.5	Threonine	0.3
DE	12.0	Starch + Sugars	48.0	Av Phos	-	Arginine	0.5
Crude Fibre	2.0	FME	12.5	Magnesium	0.1	PDIA	-
Oil (EE)	1.8	ERDP (@ 2)	-	Potassium	0.5	PDIN	-
Oil (AH)	2.7	ERDP (@ 5)	-	Sodium	0.05	PDIE	-
EFA	2.0	ERDP (@ 8)	7.2	Chloride	0.05	Met DI	-
Ash	1.5	DUP (@ 2)	-	Total Lysine	0.4	Lys DI	-

Starch 1.8% NDF 38% Other 8.2%

Sugars 3% Ash 1.5%

Protein 11.6% Oil 2.7%

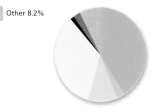

Oilseeds and By-Products

Introduction

By-product of oil manufacture, obtained by extraction of partially decorticated seeds of safflower Carthamus tinctorius L.

The Safflower was historically grown as an oilseed group in USA, Central America and India, but is now in decline. The oil (saturated) is removed by expeller, but the resultant meal contains many hulls due to poor decortication.

Origin/Manufacture

Not widely available.

Nutritional Benefit

The protein is very digestible, high in methionine, and cystine but low in lysine.

Colour/Texture

Meal or cake form.

Palatability

Unpalatable.

Limits to Usage (Anti-Nutritional Factors)

Product containing hulls is not suitable for ruminants, as the hulls are undigestible. Even in non-ruminants, its taste must be masked by other components. If decorticated, it contains a useful protein level.

Phytic acids present can bind minerals and make the meal bitter.

Concentrate Inclusion % per species

	Inc %		Inc %		Inc %
Calf	5	Creep	0	Chick	0
Dairy	10	Weaner	5	Broiler	0
Beef	10	Grower	5	Breeder	5
Lamb	5	Finisher	7.5	Layer	5
Ewe	10	Sow	7.5		

Storage/Processing

Can go rancid due to oil levels

Alternative Names

Bulk Density

Typical Analysis

Dry Matter	91.0	NCDG	-	DUP (@ 8)	-	Meth & Cysteine	-
Crude Protein	19.0	NDF	53.5	Salt	0.05	Tryptophan	-
DCP	15.0	ADF	-	Ca	0.25	Threonine	-
MER	5.5	Starch	0	Total Phos	0	Arginine	-
MEP	-	Sugar	2.0	Av Phos	0	PDIA	-
DE	-	Starch + Sugars	-	Magnesium	0.6	PDIN	-
Crude Fibre	40.0	FME	-	Potassium	-	PDIE	-
Oil (EE)	0.4	ERDP (@ 2)	-	Sodium	0.05	Met DI	-
Oil (AH)	0.5	ERDP (@ 5)	-	Chloride	-	Lys DI	-
EFA	-	ERDP (@ 8)	-	Total Lysine	0		
Linoleic	-	DUP (@ 2)	-	Avail Lysine	0		
Ash	5.0	DUP (@ 5)	-	Methionine	-		

- Starch 0%
- Sugars 2%
- Protein 19%
- NDF 53.5%
- Ash 5%
- Oil 0.5%
- Other 20%

Oilseeds and By-Products

Introduction

Produced from the nuts of the Sal Tree which is widely grown in India and South East Asia. The kernel of the nut contains about 10-15% oil, which is removed and used locally in food or sold to the worldwide confectionery industry. The remaining product is either dumped or dried for local use and/or exported.

Origin/Manufacture

Asia, and processed in Europe.

Nutritional Benefit

Low nutritive value and used in relatively small amounts.

Colour/Texture

Pale yellow.brown meal or pellets.

Palatability

Poor.

Limits to Usage (Anti-Nutritional Factors)

It contains high levels of tannins, which are unpalatable and interact with other components of the ration, reducing nutritional availability. The use of ammonia, caustic soda or other alkalis will reduce tannin levels considerably.

Concentrate Inclusion % per species

	Inc %		Inc %		Inc %
Calf	0	Creep	0	Chick	0
Dairy	2.5	Weaner	0	Broiler	0
Beef	5.0	Grower	0	Breeder	0
Lamb	0	Finisher	0	Layer	0
Ewe	2	Sow	0		

Storage/Processing

Alternative Names

Bulk Density

Typical Analysis

Dry Matter	90.0	NCGD	55.0	DUP (@ 5)	2.6	Avail Lysine	0.3
Crude Protein	9.0	NDF	31.2	DUP (@ 8)	3.3	Methionine	0.6
DCP	6.0	ADF	20.0	Salt	0.05	Meth & Cysteine	0.8
MER	8.2	Starch	26.0	Ca	0.3	Tryptophan	0.3
MEP	8.1	Sugar	27.5	Total Phos	0.25	Threonine	0.6
DE	8.4	Starch + Sugars	53.5	Av Phos	0.1	Arginine	0.5
Crude Fibre	4.2	FME	7.7	Magnesium	0.2	PDIA	-
Oil (EE)	2.3	ERDP (@ 2)	6.7	Potassium	1.0	PDIN	-
Oil (AH)	2.8	ERDP (@ 5)	4.9	Sodium	0.02	PDIE	-
EFA	0.5	ERDP (@ 8)	4.4	Chloride	0.05	Met DI	-
Ash	3.5	DUP (@ 2)	1.2	Total Lysine	0.6	Lys DI	-

Starch 26% NDF 31.2% Other 0%
Sugars 27.5% Ash 3.5%
Protein 9% Oil 2.8%

Oilseeds and By-Products

Introduction

By-product of oil manufacture, obtained by pressing of seeds of the sesame plant Sesanum indicum L.

The sesame plant stands erect, with small pods which are easily harvested. It is grown for its oil (polyunsaturated) which is expelled or extruded.

Origin/Manufacture

S. America, S. Europe, Africa, Middle East and Far East.

Nutritional Benefit

The meal/cake is high in protein, low in fibre, and has an average energy level. The protein is rich in Methionine and Arginine but low in lysine. For ruminants, the protein is 30% undegradable in the rumen. It is normally only used in adult animal rations.

Colour/Texture

Tan brown meal or cake.

Palatability

Reasonable.

Limits to Usage (Anti-Nutritional Factors)

Can have a laxative effect at higher levels. Also where oil is present, soft carcass fat can result and in dairy, it may taint milk. In poultry, the phytic acid present may bind phosphorus and extra calcium may also be needed.

Concentrate Inclusion % per species

	Inc %		Inc %		Inc %
Calf	0	Creep	0	Chick	0
Dairy	10	Weaner	0	Broiler	0
Beef	10	Grower	5	Breeder	5
Lamb	0	Finisher	5	Layer	5
Ewe	10	Sow	6		

Storage/Processing

May go rancid if stored for a long time. Rancid meal will affect Vitamin E availability.

Alternative Names

Bulk Density

432 - 480 Kg/m³

Typical Analysis

Dry Matter	9.0	NCGD	66.3	DUP (@ 5)	-	Avail Lysine	1.0
Crude Protein	42.0	NDF	-	DUP (@ 8)	-	Methionine	0.82
DCP	38.0	ADF	-	Salt	0.1	Meth & Cysteine	1.6
MER	12.3	Starch	-	Ca	2.1	Tryptophan	5.0
MEP	10.6	Sugar	-	Total Phos	1.4	Threonine	0.6
DE	13.1	Starch + Sugars	-	Av Phos	0.5	Arginine	5.0
Crude Fibre	7.0	FME	-	Magnesium	0.7	PDIA	-
Oil (EE)	9.3	ERDP (@ 2)	-	Potassium	1.2	PDIN	-
Oil (AH)	10.5	ERDP (@ 5)	-	Sodium	0.05	PDIE	-
EFA	-	ERDP (@ 8)	-	Chloride	0.05	Met DI	-
Ash	11.8	DUP (@ 2)	-	Total Lysine	1.3	Lys DI	-

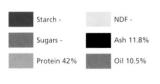

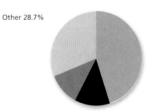

- Starch -
- NDF -
- Other 28.7%
- Sugars -
- Ash 11.8%
- Protein 42%
- Oil 10.5%

Oilseeds and By-Products

Introduction
Shea nuts are grown for their edible oil which is used in confectionery. The oil is removed by expellers (40 - 45% oil).

Origin/Place of Manufacture
Asia, Africa.

Nutritional Benefit
Shea nuts are of low feed value and poorly digested.

Colour/Texture
Yellow brown meal/cake.

Palatability
Bitter.

Limits to Usage (Anti-Nutritional Factors)
Contains saponine which can damage the gut lining and cause haemolysis of blood cells. The saponine can be removed by heat treatment.

Concentrate Inclusion % per species

	Inc %		Inc %		Inc %
Calf	0	Creep	0	Chick	0
Dairy	5	Weaner	0	Broiler	0
Beef	7.5	Grower	0	Breeder	0
Lamb	0	Finisher	0	Layer	0
Ewe	5	Sow	0		

Storage/Processing

Alternative Names

Bulk Density

Typical Analysis

Dry Matter	88.0	NCGD	68.0	DUP (@ 5)	4.5	Avail Lysine	-
Crude Protein	16.5	NDF	35.7	DUP (@ 8)	5.3	Methionine	-
DCP	3.0	ADF	-	Salt	0.15	Meth & Cysteine	-
MER	6.8	Starch	0.2	Ca	0.25	Tryptophan	-
MEP	-	Sugar	7.0	Total Phos	0.25	Threonine	-
DE	15.3	Starch + Sugars	7.2	Av Phos	0.1	Arginine	-
Crude Fibre	8.7	FME	4.5	Magnesium	0.2	PDIA	7.9
Oil (EE)	6.5	ERDP (@ 2)	8.6	Potassium	2.0	PDIN	11.3
Oil (AH)	7.9	ERDP (@ 5)	6.8	Sodium	0.05	PDIE	-
EFA	6.0	ERDP (@ 8)	5.9	Chloride	0.2	Met DI	-
Ash	6.9	DUP (@ 2)	4.0	Total Lysine	-	Lys DI	-

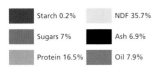

Starch 0.2% — NDF 35.7% — Other 17.1%
Sugars 7% — Ash 6.9%
Protein 16.5% — Oil 7.9%

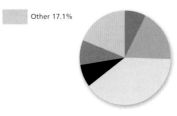

| Miscellaneous | |

Introduction

Sodium bicarbonate is used in ruminant diets to buffer the rumen while in monogastrics it is used as a sodium source.

Origin/Place of Manufacture

UK and around the world.

Nutritional Benefit

Ideal to stabilise the rumen pH. Tends to increase butterfat by increasing the buffering capacity. Used in poultry to provide sodium without chloride thereby reducing potential for wet litter.

Colour/Texture

White powder.

Palatability

If used heavily, spices may be required to encourage intake.

Limits to Usage (Anti-nutritional factors)

Concentrate Inclusion % per species

	Inc %		Inc %		Inc %
Calf	1	Creep	0	Chick	0.1
Dairy	2	Weaner	0	Broiler	0.2
Beef	1	Grower	0	Breeder	0.1
Lamb	1	Finisher	0	Layer	0.1
Ewe	1	Sow	0		

Storage/Processing

Alternative Names

Bulk Density

Typical Analysis

Dry Matter	98.0	NDF	0	DUP (@ 8)	0	Meth & Cysteine	0
Crude Protein	0	ADF	0	Salt	-	Tryptophan	0
DCP	0	Starch	0	Ca	0	Threonine	0
MER	0	RRS	0	Total Phos	0	Arginine	0
MEP	0	Sugar	0	Av Phos	0	PDIA	0
DE	0	Starch + Sugars	0	Magnesium	0	PDIN	0
Crude Fibre	0	FME	0	Potassium	0	PDIE	0
Oil (EE)	0	ERDP (@ 2)	0	Sodium	28.0	Met DI	0
Oil (AH)	0	ERDP (@ 5)	0	Chloride	0	Lys DI	0
EFA	0	ERDP (@ 8)	0	Total Lysine	0		
Ash	60.0	DUP (@ 2)	0	Avail Lysine	0		
NCDG	0	DUP (@ 5)	0	Methionine	0		

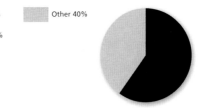

- Starch 0%
- Sugars 0%
- Protein 0%
- NDF 0%
- Ash 60%
- Oil 0%
- Other 40%

Cereals and By-Products

Introduction

Grains of Sorghum bicolor (L.) Moench s.i.

One of the most widely used straight feed grains in the United States of America. Most sorghum by-products from processing are reportedly unpalatable and not widely used.

Origin/Manufacture

USA, Africa, China, India, Pakistan.

Nutritional Benefit

The grain is mainly used for feed, although as a grass the total plant could be used as forage. The seeds are low in fibre and nutritionally similar to maize grain, but lacking in the xanthophylls required for layers rations.

Colour/Texture

Dark brown.

Palatability

Relatively palatable. High tannin level could reduce intake.

Limits to Usage (Anti-Nutritional Factors)

Dark brown or purple seeds contain a lot of tannin, and will reduce protein digestibility. Not ideal for wet feeding systems. White seeds contain little tannin and are an ideal feed, although they can have a constipating effect. NB. It is essential that the type of Sorghum is known to determine the true nutritional value.

Concentrate Inclusion % per species

	Inc %		Inc %		Inc %
Calf	5	Creep	0	Chick	0
Dairy	10	Weaner	0	Broiler	0
Beef	10	Grower	5	Breeder	5
Lamb	5	Finisher	5	Layer	5
Ewe	10	Sow	6		

Storage/Processing

The whole grain should be processed for most ruminants except sheep (grain coat reduces digestion otherwise). Processing removes the seed coat, which improves digestion. It should not be over processed, ie. too finely ground, or intake will be reduced.

Alternative Names

Milo

Bulk Density

Seeds 500 - 565 Kg/m³   **Meal** 475 - 500 Kg/m³

Typical Analysis

Dry Matter	85.0	NCDG	93.0	DUP (@ 5)	6.3	Avail Lysine	0.15
Crude Protein	10.9	NDF	13.1	DUP (@ 8)	6.5	Methionine	0.2
DCP	7.3	ADF	5.4	Salt	0.15	Meth & Cysteine	0.4
MER	13.5	Starch	68.3	Ca	0.05	Tryptophan	0.1
MEP	15.6	Sugar	2.0	Total Phos	0.5	Threonine	0.4
DE	16.0	Starch + Sugars	82.5	Av Phos	0.15	Arginine	0.4
Crude Fibre	3.0	FME	12.6	Magnesium	0.2	PDIA	6.5
Oil (EE)	3.5	ERDP (@ 2)	4.4	Potassium	0.4	PDIN	8.7
Oil (AH)	3.7	ERDP (@ 5)	3.1	Sodium	0.3	PDIE	14.0
EFA	1.7	ERDP (@ 8)	2.8	Chloride	0.1	Met DI	-
Ash	2.0	DUP (@ 2)	5.5	Total Lysine	0.25	Lys DI	-

- Starch 68.3%
- Sugars 2%
- Protein 10.9%
- NDF 13.1%
- Ash 2%
- Oil 3.7%
- Other 0%

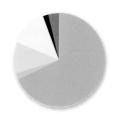

Oilseeds and By-Products

Introduction

By-product of oil manufacture, obtained from soya beans after extraction and appropriate heat treatment.

Soya beans when crushed produce three main products: oil, meal and hulls. The remainder is then heated and flaked. The oil is extracted (or occasionally expelled) and the remainder toasted to reduce anti-nutritional factors (ANFs) and drive off remaining solvents. For every tonne of soya beans crushed approximately 750 kgs of soya bean meal are produced. Northern hemisphere beans are harvested in Oct. Nov. and Southern hemisphere Feb. Apr.

Origin/Place of Manufacture

USA, Brazil, with beans imported for UK processing.

Nutritional Benefit

Soya bean meal is probably the best quality vegetable protein source widely used around the world. It is high in protein and energy and has a good amino acid profile, being high in lysine, although methionine is low. 95% of the nitrogen present is true protein, making it ideal for all livestock. High in phosphorus of which 50% is available. Hipro Soya does not have the hulls re-blended and is, therefore, lower in fibre, but higher in protein than other sources. A good source of B and D vitamins.

Colour/Texture

Pale yellow textured meal/pellets.

Palatability

Good.

Limits to Usage (Anti-Nutritional Factors)

Trypsin inhibitors and haemaggluttins, are usually inactivated by the heat treatment during manufacture. Other inactivated ANF's could include goitrogens, saponins, oestrogens, anti-vitamins and phytates.

Concentrate Inclusion % per species

	Inc %		Inc %		Inc %
Calf	20	Creep	20	Chick	25
Dairy	35	Weaner	25	Broiler	30
Beef	35	Grower	30	Breeder	35
Lamb	20	Finisher	30	Layer	35
Ewe	30	Sow	30		

Storage/Processing

Alternative Names

Bulk Density

Exp Meal 575 - 645 Kg/m³   **Ext 44** 550 - 610 Kg/m³
Ext 50 650 - 675 Kg/m³   **Whole Soya Bean** 720 - 800 Kg/m³

Typical Analysis	Soya Hipro	Soya Lopro	Typical Analysis	Soya Hipro	Soya Lopro
Dry Matter	90.0	88.0	DUP (@ 5)	14.0	11.9
Crude Protein	55.0	47.0	DUP (@ 8)	24.5	11.3
DCP	52.0	43.0	Salt	0.05	0.1
MER	13.6	12.9	Ca	0.4	0.3
MEP	12.0	10.7	Total Phos	0.85	0.7
DE	15.5	14.8	Av Phos	0.7	0.6
Crude Fibre	4.0	8.2	Magnesium	0.3	0.3
Oil (EE)	2.4	2.0	Potassium	2.0	2.0
Oil (AH)	2.6	2.3	Sodium	0.01	0.01
EFA	0.9	0.9	Chloride	0.02	0.03
Ash	6.5	7.2	Total Lysine	3.1	3.0
NCGD	93.0	84.2	Avail Lysine	2.9	2.8
NDF	11.0	16.1	Methionine	0.7	0.7
ADF	6.0	8.0	Meth & Cysteine	1.5	1.5
Starch	5.0	4.5	Tryptophan	0.9	0.8
Sugar	11.0	10.0	Threonine	2.0	1.9
Starch + Sugars	16.0	14.5	Arginine	4.2	3.6
FME	12.7	12.4	PDIA	21.0	18.9
ERDP (@ 2)	42.1	38.2	PDIN	40.0	34.8
ERDP (@ 5)	32.5	34.9	PDIE	27.1	25.0
ERDP (@ 8)	26.5	32.5	Met DI	0.4	0.4
DUP (@ 2)	10.6	41.5	Lys DI	1.9	1.8

Brazilian Soya

Typical Analysis

Dry Matter	88.0	NCGD	77.8	DUP (@ 5)	9.0	Avail Lysine	3.0
Crude Protein	51.0	NDF	13.4	DUP (@ 8)	11.9	Methionine	0.65
DCP	50.0	ADF	9.2	Salt	0.07	Meth & Cysteine	1.30
MER	13.3	Starch	4.5	Ca	0.3	Tryptophan	0.6
MEP	11.1	Sugar	10.0	Total Phos	0.65	Threonine	2.0
DE	15.7	Starch + Sugars	14.5	Av Phos	0.20	Arginine	3.5
Crude Fibre	7.0	FME	12.0	Magnesium	0.30	PDIA	19.3
Oil (EE)	2.0	ERDP (@ 2)	41.0	Potassium	2.3	PDIN	36.2
Oil (AH)	2.3	ERDP (@ 5)	35.0	Sodium	0.01	PDIE	25.1
EFA	1.0	ERDP (@ 8)	33.0	Chloride	0.03	Met DI	0.35
Ash	6.7	DUP (@ 2)	4.5	Total Lysine	3.3	Lys DI	1.8

Starch 5% NDF 13% Other 7%

Sugars 11% Ash 6.5%

Protein 55% Oil 2.5%

Oilseeds and By-Products

Introduction

Oil obtained from soya beans.

Beans are dehulled, heated, flaked and the oil extracted by solvent extraction. After desolventising and degumming, the oil is available crude for animal feed, or further refined for human food by alkali refining and deodorisation. For every tonne of soya beans crushed approximately 190 kgs of oil are produced.

Origin/Place of Manufacture

UK, Europe, USA, S. America.

Nutritional Benefit

Rich in linoleic acid (C18:2) and energy, it is ideal for all non-ruminant rations. Cheaper by-products are usually fed in commercial situations, eg. Soya Acid Oil.

Colour/Texture

Orange/brown, translucent liquid.

Palatability

Limits to Usage (Anti-Nutritional Factors)

Concentrate Inclusion % per species

	Inc %		Inc %		Inc %
Calf	2.5	Creep	2.5	Chick	2.5
Dairy	2.5	Weaner	2.5	Broiler	5
Beef	2.5	Grower	2.5	Breeder	5
Lamb	2.5	Finisher	2.5	Layer	2.5
Ewe	2.5	Sow	1.0		

Storage/Processing

Can be stored in mild steel or fibre glass tanks.

Alternative Names

Soya Oil.

Bulk Density

Typical Analysis

Dry Matter	99.5	NCGD	93.0	DUP (@ 8)	0	Meth & Cysteine	0
Crude Protein	0	NDF	0	Salt	0	Tryptophan	0
DCP	0	ADF	0	Ca	0	Threonine	0
MER	36.0	Starch	0	Total Phos	0	Arginine	0
MEP	37.0	Sugar	0	Av Phos	0	PDIA	0
DE	40.5	Starch + Sugars	0	Magnesium	0	PDIN	0
Crude Fibre	0	FME	1.4	Potassium	0	PDIE	0
Oil (EE)	98.7	ERDP (@ 2)	0	Sodium	0	Met DI	0
Oil (AH)	98.7	ERDP (@ 5)	0	Chloride	0	Lys DI	0
EFA	60.0	ERDP (@ 8)	0	Total Lysine	0		
Linoleic	52.0	DUP (@ 2)	0	Avail Lysine	0		
Ash	0.5	DUP (@ 5)	0	Methionine	0		

Starch 0%	NDF 0%	Other 0.8%
Sugars 0%	Ash 0.5%	
Protein 0%	Oil 98.7%	

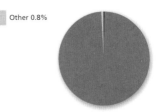

Oilseeds and By-Products

Introduction

Soya Beans Glycine max. L. Merr. subjected to an appropriate heat treatment.

Soya is the most important protein source in the world. The soya seeds are de-hulled and heat treated either by steam followed, potentially, by extrusion, toasting, micronising or jet sploding to produce a high oil, high protein product. Used commonly in young animal rations, its usage depends on the combined cost of soya meal and soya bean oil. The heat from processing reduces the levels of anti-nutritional factors (ANFs) which would otherwise reduce protein digestion. It also breaks down the plant cell wall, making it more easily digested.

Origin/Place of Manufacture

UK, worldwide.

Palatability

Excellent.

Nutritional Benefit

Ideal for all rations, especially young animals due to palatability, high energy and protein (40%) level. It is high in essential fatty acids, making it ideal for pig and poultry feed. The oil present may have some degree of rumen protection and could, therefore, enhance milk yield and modify quality. The heat during manufacture increases the level of undegradable protein.

Colour/Texture

Pale golden yellow.

Palatability

Excellent.

Limits to Usage (Anti-Nutritional Factors)

ANFs, eg. Trypsin and haemagluttin inhibitors, are removed by heating/cooking.
Finishing beef animals should not be fed high levels or soft fat will result.
It may also increase butterfat levels in milk.

Concentrate Inclusion % per species

	Inc %		Inc %		Inc %
Calf	10	Creep	20	Chick	20
Dairy	15	Weaner	20	Broiler	25
Beef	15	Grower	15	Breeder	20
Lamb	10	Finisher	10	Layer	20
Ewe	15	Sow	10		

Storage/Processing

Higher oil content will increase risk of rancidity if it is stored for long periods.

Alternative Names

Bulk Density

Typical Analysis

Dry Matter	89.0	NCDG	8.9	DUP (@ 5)	6.6	Avail Lysine	2.4
Crude Protein	40.0	NDF	12.3	DUP (@ 8)	8.9	Methionine	0.6
DCP	38.0	ADF	8.2	Salt	0.1	Meth & Cysteine	1.1
MER	16.1	Starch	3.7	Ca	0.3	Tryptophan	0.6
MEP	16.9	Sugar	8.3	Total Phos	0.5	Threonine	1.5
DE	19.3	Starch + Sugars	12.0	Av Phos	0.25	Arginine	0.5
Crude Fibre	6.0	FME	8.5	Magnesium	0.25	PDIA	18.4
Oil (EE)	20.5	ERDP (@ 2)	32.0	Potassium	1.7	PDIN	27.8
Oil (AH)	20.7	ERDP (@ 5)	27.9	Sodium	0.02	PDIE	21.2
EFA	10.9	ERDP (@ 8)	26.0	Chloride	0.05	Met DI	0.4
Ash	5.5	DUP (@ 2)	3.4	Total Lysine	2.8	Lys DI	1.1

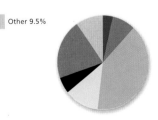

- Starch 3.7%
- NDF 12.3%
- Other 9.5%
- Sugars 8.3%
- Ash 5.5%
- Protein 40%
- Oil 20.7%

Oilseeds and By-Products

Introduction

By-product obtained during dehulling of soya beans.

Soya Beans are normally dehulled prior to crushing and the resulting hulls are either sold as a meal or pelleted. Widely used in the USA.

Origin/Manufacture

Europe, USA, Asia.

Nutritional Benefit

A good source of digestible fibre, with average protein and reasonable energy levels.

Colour/Texture

Brown/tan flakes or pellets.

Palatability

Average.

Limits to usage

High fibre level may limit intake.

Concentrate Inclusion % per species

	Inc %		Inc %		Inc %
Calf	10	Creep	0	Chick	0
Dairy	25	Weaner	0	Broiler	0
Beef	25	Grower	0	Breeder	0
Lamb	10	Finisher	5	Layer	0
Ewe	20	Sow	10		

Storage/Processing

Alternative Names

Bulk Density

Typical Analysis

Dry Matter	90.0	NCDG	76.0	DUP (@ 5)	2.9	Avail Lysine	0.5
Crude Protein	11.6	NDF	67.5	DUP (@ 8)	6.2	Methionine	0.2
DCP	6.5	ADF	30.5	Salt	0.05	Meth & Cysteine	0.35
MER	11.9	Starch	5.0	Ca	0.5	Tryptophan	0.1
MEP	3.7	Sugar	9.0	Total Phos	0.2	Threonine	0.8
DE	8.2	Starch + Sugars	14.0	Av Phos	0.1	Arginine	0.8
Crude Fibre	35.0	FME	9.6	Magnesium	0.25	PDIA	4.3
Oil (EE)	1.8	ERDP (@ 2)	8.0	Potassium	1.0	PDIN	7.4
Oil (AH)	2.4	ERDP (@ 5)	6.8	Sodium	0.02	PDIE	10.8
EFA	1.2	ERDP (@ 8)	6.1	Chloride	0.02	Met DI	0.18
Ash	4.5	DUP (@ 2)	1.9	Total Lysine	0.7	Lys DI	0.89

Starch 5% NDF 67.5% Other 0%

Sugars 9% Ash 4.5%

Protein 11.6% Oil 2.4%

Cereals and By-Products

Introduction

This is the liquid remaining after maize or wheat fermentation, containing spent yeast and unfermented soluble components. The liquid is evaporated to further concentrate the spent wash syrup.

Origin/Place of Manufacture

Scotland.

Nutritional Benefit

Good protein source.

Colour/Texture

Golden brown, extremely viscous, liquid. Consistency will depend on the level of fine solids present.

Palatability

Very palatable.

Limits to Usage (Anti-Nutritional Factors)

pH low 3.5 - 4.0. High in potassium which may cause scouring.

Concentrate Inclusion % per species

	Inc %		Inc %		Inc %
Calf	0	Creep	0	Chick	0
Dairy	10	Weaner	0	Broiler	0
Beef	10	Grower	0	Breeder	0
Lamb	0	Finisher	5	Layer	0
Ewe	5	Sow	5		

Storage/Processing

The material is highly prone to gelling.

Alternative Names

Grain Syrup.

Bulk Density

1050 - 1150 Kg/m³

Typical Analysis

Dry Matter	40.0	NCGD	-	DUP (@ 5)	-	Avail Lysine	-
Crude Protein	21.5	NDF	-	DUP (@ 8)	-	Methionine	-
DCP	-	ADF	-	Salt	-	Meth & Cysteine	-
MER	13.5	Starch	-	Ca	0.03	Tryptophan	-
MEP	-	Sugar	-	Total Phos	0.1	Threonine	-
DE	9.2	Starch + Sugars	-	Av Phos	-	Arginine	-
Crude Fibre	-	FME	-	Magnesium	0.7	PDIA	-
Oil (EE)	-	ERDP (@ 2)	-	Potassium	-	PDIN	-
Oil (AH)	-	ERDP (@ 5)	-	Sodium	0.03	PDIE	-
EFA	-	ERDP (@ 8)	-	Chloride	-	Met DI	-
Ash	12.0	DUP (@ 2)	-	Total Lysine	-	Lys DI	-

Starch -   NDF -   Other 66.5%

Sugars -   Ash 12%

Protein 21.5%   Oil -

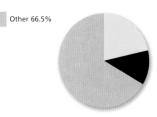

Forages and Stock Feeds

Introduction

The fibrous stalks from growing barley, wheat, oats or legumes. Straw consists of the stems and leaves of the cereals after the removal of the ripe seed.

Origin/Place of Manufacture

UK and worldwide.

Nutritional Benefit

A good source of long fibre but low in protein and energy. Oat straw is often claimed to be the most palatable. The composition of the straw is influenced more by stage of maturity of the crop at harvesting and environment than by the variety grown. The crude protein content is low (2-5%) and of low digestibility. Straw is made up of 40-45% cellulose, 30-50% hemicellulose and 8-12% lignin. The straw of pea and bean is richer in protein, calcium and magnesium than cereal straw. The thick, fibrous stalks mean they are more difficult to dry and easily get mouldy compared to cereal straw.

Colour/Texture

Pale yellow for cereal, and green for legumes.

Palatability

Often needs to be mixed or coated with molasses to encourage intake.

Limits to Usage (Anti-Nutritional Factors)

Energy levels (6-7 MJ/Kg DM) limit inclusion rates in ruminant ration. Its fibrous nature means it is unsuitable for non-ruminants.

Forage Inclusion % per species

	Inc %		Inc %		Inc %
Calf	5	Creep	0	Chick	0
Dairy	10	Weaner	0	Broiler	0
Beef	10	Grower	0	Breeder	0
Lamb	3	Finisher	0	Layer	0
Ewe	0	Sow	0		

Storage/Processing

The nutritional value can be improved by treatment with alkalis, eg. caustic soda and ammonium hydroxide.

Alternative Names

Bulk Density

Barley Straw

Typical Analysis

Dry Matter	87.0	NCDG	40.0	DUP (@ 5)	1.7	Avail Lysine	-
Crude Protein	4.0	NDF	84.4	DUP (@ 8)	1.8	Methionine	-
DCP	1.0	ADF	51.0	Salt	0.5	Meth & Cysteine	-
MER	6.5	Starch	1.0	Ca	0.5	Tryptophan	-
MEP	-	Sugar	2.0	Total Phos	0.15	Threonine	-
DE	-	Starch + Sugars	3.0	Av Phos	-	Arginine	-
Crude Fibre	44.0	FME	6.0	Magnesium	0.05	PDIA	1.25
Oil (EE)	1.2	ERDP (@ 2)	1.6	Potassium	0.1	PDIN	2.5
Oil (AH)	1.6	ERDP (@ 5)	1.4	Sodium	1.0	PDIE	4.75
EFA	-	ERDP (@ 8)	1.3	Chloride	0.1	Met DI	-
Ash	7.0	DUP (@ 2)	1.4	Total Lysine	-	Lys DI	-

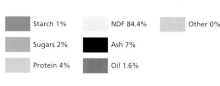

Starch 1%	NDF 84.4%	Other 0%
Sugars 2%	Ash 7%	
Protein 4%	Oil 1.6%	

Forages and Stock Feeds

Introduction

Wheat or barley straw is first ground, hammer milled and then treated with sodium hydroxide (caustic soda) or ammonia to make the carbohydrates more available.

Origin/Place of Manufacture

UK.

Nutritional Benefit

A good source of digestible fibre but very low in protein and other nutrients.
Ideal to extend forage or complement starch source. Mineral and vitamin levels are extremely low except for sodium which is supplied when caustic treated. Ammonia treatment raises the protein level to 7% by supplying non-protein nitrogen.

Colour/Texture

Yellow/brown pellets.

Palatability

Good.

Limits to Usage (Anti-Nutritional Factors)

Sodium content means adequate water should be made available. Avoid high intakes for prolonged periods as this can lead to alkalosis.

Foarge Inclusion % per species

	Inc %		Inc %		Inc %
Calf	5	Creep	0	Chick	0
Dairy	15	Weaner	0	Broiler	0
Beef	15	Grower	0	Breeder	0
Lamb	5	Finisher	0	Layer	0
Ewe	10	Sow	5		

Storage/Processing

Stores well.

Alternative Names

Caustic Straw.

Bulk Density

Straw - Caustic treated

Typical Analysis

Dry Matter	86.0	NDF	78.0	DUP (@ 8)	1.0	Methionine	0.02
Crude Protein	4.0	ADF	45.0	Salt	0.5	Meth & Cysteine	0.07
DCP	1.0	Starch	1.0	Ca	0.3	Tryptophan	0.1
MER	7.5	Sugar	1.5	Total Phos	0.15	Threonine	0.3
MEP	-	Starch + Sugars	2.5	Av Phos	-	Arginine	-
DE	4.0	FME	7.1	Magnesium	0.1	PDIA	1.7
Fibre	45.0	ERDP (@ 2)	5.0	Potassium	1.0	PDIN	4.0
Oil (EE)	1.3	ERDP (@ 5)	4.5	Sodium	0.05	PDIE	6.0
Oil (AH)	1.5	ERDP (@ 8)	4.1	Chlorine	0.65	Met DI	-
Ash	14.0	DUP (@ 2)	0.05	Total Lysine	0.1	Lys DI	-
NCGD	45.0	DUP (@ 5)	0.5	Avail Lysine	-		

Ammonia Treated

Typical Analysis

Dry Matter	87.0	NCGD	-	DUP (@ 5)	-	Avail Lysine	-
Crude Protein	7.0	NDF	78.2	DUP (@ 8)	-	Methionine	-
DCP	-	ADF	48.3	Salt	0.25	Meth & Cysteine	-
MER	7.8	Starch	1.0	Ca	0.4	Tryptophan	-
MEP	-	Sugar	2.0	Total Phos	0.1	Threonine	-
DE	-	Starch + Sugars	3.0	Av Phos	-	Arginine	-
Fibre	42.0	FME	6.8	Magnesium	0.85	PDIA	-
Oil (EE)	2.0	ERDP (@ 2)	-	Potassium	-	PDIN	-
Oil (AH)	2.4	ERDP (@ 5)	-	Sodium	0.1	PDIE	-
EFA	-	ERDP (@ 8)	-	Chloride	-	Met DI	-
Ash	5.5	DUP (@ 2)	-	Total Lysine	-	Lys DI	-

- Starch 1%
- Sugars 1.5%
- Protein 4%
- NDF 78%
- Ash 14%
- Oil 1.5%
- Other 0%

CONTEXT

Roots, Fruits and By-Products

Introduction

Molassed Sugar Beet Pulp: By-product of the manufacture of sugar comprising dried sugar-beet pulp, to which molasses has been added.

Sugar beet root grown to produce sugar. It is crushed and extracted to remove sugars. Sugar beet pulp contains the remaining fibre, tops and roots, processing residues and beet molasses. Ideal for ruminants and provides a rumen buffer.

Origin/Place of Manufacture UK and other parts of Europe, USA.

Nutritional Benefit

High in rumen fermentable energy (FME) in a palatable form. Similar in energy levels to barley but low/average in terms of protein. Contains highly digestible fibre (mainly cellulose) which is suited to ruminants as it maintains rumen condition and encourages acetate production. However, it contains too much fibre for young pigs and poultry. Ideally it should be soaked before feeding to horses. It has low phosphorus and starch levels, but good levels of sugars. It has a high liquid absorbency and can, therefore, be used as a silage additive to retain the feed value from effluent.

Colour/Texture

Brownish grey pellets or shreds.

Palatability

Very good.

Limits to Usage (Anti-Nutritional Factors)

Slightly laxative. Its bulk nature, lower density and fibrous nature make it unsuitable for poultry.

Concentrate Inclusion % per species

	Inc %		Inc %		Inc %
Calf	20	Creep	0	Chick	0
Dairy	30	Weaner	10.0	Broiler	0
Beef	40	Grower	10.0	Breeder	0
Lamb	20	Finisher	15.0	Layer	0
Ewe	50	Sow	25.0		

Storage/Processing

Will absorb large quantities of water if not stored correctly.

Alternative Names Beet Pulp.

Bulk Density

Ground 550 Kg/m³   Unground 240 - 288 Kg/m³

Typical Analysis	Sugar Beet Molassed	Sugar Beet Unmolassed	Typical Analysis	Sugar Beet Molassed	Sugar Beet Unmolassed
Dry Matter	90.0	90.0	DUP (@ 5)	3.8	3.3
Crude Protein	11.0	10.0	DUP (@ 8)	4.5	3.8
DCP	7.2	7.2	Salt	0.5	0.2
MER	12.5	12.5	Ca	0.95	0.8
MEP	5.8	3.5	Total Phos	0.15	0.1
DE	13.1	13.2	Av Phos	0.05	0.3
Crude Fibre	15.0	19.0	Magnesium	0.15	0.25
Oil (EE)	0.4	0.4	Potassium	2.0	1.2
Oil (AH)	0.4	0.5	Sodium	0.5	0.4
EFA	0.15	0.15	Chloride	0.25	0.11
Ash	8.5	6.8	Total Lysine	0.5	0.4
NCDG	84.0	87.0	Avail Lysine	0.35	0.3
NDF	30.5	35.0	Methionine	0.1	0.1
ADF	18.0	22.0	Meth & Cysteine	0.2	0.25
Starch	2.0	1.5	Tryptophan	0.03	0.05
Sugar	23.0	6.5	Threonine	0.32	0.3
Starch + Sugars	25.0	8.0	Arginine	0.25	0.26
FME	12.3	12.4	PDIA	3.9	3.9
ERDP (@ 2)	6.4	6.9	PDIN	6.5	6.5
ERDP (@ 5)	4.9	5.5	PDIE	10.6	10.7
ERDP (@ 8)	4.5	4.6	Met DI	0.1	0.1
DUP (@ 2)	2.6	1.8	Lys DI	0.5	0.4

Starch 2%   NDF 30.5%   Other 24.6%

Sugars 23%   Ash 8.5%

Protein 11%   Oil 0.4%

CONTEXT

Roots, Fruits and By-Products

Introduction

By-products of the manufacture of sugar, consisting of extracted and dried pieces of sugar-beet Beta Vulagaris L. ssp. Vulgaris var. altissima Doell.

This is the pressed fibre from sugar beet processing after hot water extraction. It is pressed, but not dried.

Origin/Place of Manufacture

UK.

Nutritional Benefit

A palatable forage and partial concentrate extender, which can be used to replace dried sugar beet feed. Ideal for ruminants, but not pigs and poultry. With a typical DM content of 20-30%, this product is a useful energy feed for dairy cows, beef cattle and sheep. Ideal for use in complete feeds or for use as a midday feed. This buffer feed and moist concentrate is a high energy succulent to supplement silage when supplies are limited or the quality poor. Sugar Beet Pulp improves the energy density of rations based on straw and hay. In complete diet systems, its high energy density and physical nature enhance the quality of the mix.

Colour/Texture

Brown.

Palatability

Good.

Limits to Usage (Anti-Nutritional Factors)

Low in phosphorus. If fed fresh, it should be used within 5-7 days of leaving the factory to avoid the growth of harmful moulds. If kept for more than 2-3 days it should be ensiled and fully sealed. Care must be taken to ensure that there is adequate coarse long fibre in the ration. In all cases, it is advisable to use a high phosphorus mineral/vitamin supplement.

Forage Inclusion % per species

	Inc %		Inc %		Inc %
Calf	5	Creep	0	Chick	0
Dairy	30	Weaner	0	Broiler	0
Beef	30	Grower	0	Breeder	0
Lamb	5	Finisher	0	Layer	0
Ewe	20	Sow	0		

Storage/Processing

Product should always be tipped in a clean, dry area. Ensiling pressed sugar beet pulp It is essential to exclude all air and water from the silo to successfully store the product.

Alternative Names

Bulk Density

1 tonne/m³

Typical Analysis

Dry Matter	20.0	NDF	52.3	DUP (@ 8)	4.2	Methionine	0.1
Crude Protein	9.5	ADF	27.1	Salt	0.12	Meth & Cysteine	0.2
DCP	8.0	Starch	0.4	Ca	1.0	Tryptophan	0.05
MER	12.1	Sugar	4.6	Total Phos	0.2	Threonine	0.3
MEP	-	Starch + Sugars	5.0	Av Phos	-	Arginine	0.2
DE	-	FME	12.0	Magnesium	0.2	PDIA	3.5
Crude Fibre	20.0	ERDP (@ 2)	6.6	Potassium	1.0	PDIN	6.4
Oil (EE)	0.2	ERDP (@ 5)	4.8	Sodium	0.5	PDIE	10.3
Oil (AH)	0.3	ERDP (@ 8)	4.4	Chloride	0.25	Met DI	-
Ash	8.0	DUP (@ 2)	2.5	Total Lysine	0.4	Lys DI	-
NCGD	85.0	DUP (@ 5)	3.5	Avail Lysine	-		

- Starch 0.4%
- Sugars 4.6%
- Protein 9.5%
- NDF 52.3%
- Ash 8%
- Oil 0.3%
- Other 24.9%

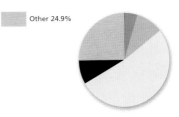

Oilseeds and By-Products

Introduction

By-product of oil manufacture obtained be expelling and/or extraction of seeds of the sunflower.

Grown for its 'soft' polyunsaturated oil, which is removed by expeller and or extraction. The seed has a typical oil content of 45% and dehulled prior to oil extraction. The hulls are added back to the product after the oil is removed.

Origin/Place of Manufacture

Sub-tropical/temperate countries.

Nutritional Benefit

Sunflower has good protein levels but a lower lysine content than soya. It is high in fibre which, if originating from the seed coat, is poorly digested and should be finely ground. Removing the hulls from the meal produces a more digestible product with higher protein level (Hipro Sunflower). Rich in sulphur-containing amino acids and phosphorus but low in lysine. Can be included in pig and poultry rations, but best suited to ruminant rations within which it can help reduce acidosis by its fibre contribution.

Colour/Texture

Grey/black meal pellets.

Palatability

Average/Good.

Limits to Usage (Anti-Nutritional Factors)

Slightly laxative at high levels.

Concentrate Inclusion % per species

	Inc %		Inc %		Inc %
Calf	2.5	Creep	0	Chick	0
Dairy	25	Weaner	0	Broiler	5
Beef	25	Grower	2.5	Breeder	10
Lamb	2.5	Finisher	5	Layer	10
Ewe	20	Sow	10		

Storage/Processing

The meal will oxidise and go rancid. Should be ground finely to break up husks.

Alternative Names

Decorticated extracted sunflower seed meal.

Bulk Density

525 - 600 Kg/m³

Typical Analysis	Standard	Hipro Suns		Typical Analysis	Standard	Hipro Suns
Dry Matter	88.0	88.0		DUP (@ 5)	3.2	4.2
Crude Protein	36.0	45.0		DUP (@ 8)	4.6	7.5
DCP	28.0	38.0		Salt	0.25	0.25
MER	9.5	11.5		Ca	0.3	0.35
MEP	7.1	8.6		Total Phos	1.2	1.3
DE	10.2	11.8		Av Phos	0.35	0.4
Crude Fibre	23	16		Magnesium	0.6	0.65
Oil (EE)	2.0	2.5		Potassium	1.2	1.2
Oil (AH)	2.5	3.0		Sodium	0.05	0.05
EFA	1.0	1.5		Chloride	3.3	3.0
Ash	7.0	6.5		Total Lysine	1.2	1.4
NCDG	60.0	74.0		Avail Lysine	1.0	1.1
NDF	47.0	42.0		Methionine	1.0	1.0
ADF	32.0	22.5		Meth & Cysteine	1.4	1.6
Starch	1.5	6.0		Tryptophan	0.5	0.55
Sugar	6.0	6.5		Threonine	1.4	1.6
Starch + Sugars	7.5	12.5		Arginine	3.0	3.3
FME	13.5	12.1		PDIA	8.4	10.0
ERDP (@ 2)	33.1	42.0		PDIN	25.2	32.0
ERDP (@ 5)	30.2	37.8		PDIE	11.7	12.9
ERDP (@ 8)	29.0	36.0		Met DI	0.25	0.32
DUP (@ 2)	1.5	1.8		Lys DI	0.7	0.85

Starch 1.5%	NDF 47%	Other 0%	
Sugars 6%	Ash 7%		
Protein 36%	Oil 2.5%		

Oilseeds and By-Products

Introduction

Seeds are dehulled, heated, flaked and the oil extracted by screw pressing (expelling) and then by solvent extraction. After desolventising and degumming, the oil is available crude for animal feeds, or further refined for human consumption.

Origin/Place of Manufacture

UK, Europe, USA, South America.

Nutritional Benefit

A soft oil with a high energy content. Rich in Linolenic acid (C18:3) and suitable for dedusting of rations.

Colour/Texture

Yellow/orange, translucent liquid.

Palatability

Average.

Limits to Usage

Inclusion % per species

	Inc %		Inc %		Inc %
Calf	2.5	Creep	2.5	Chick	2.5
Dairy	2.5	Weaner	2.5	Broiler	2.5
Beef	2.5	Grower	2.5	Breeder	2.5
Lamb	2.5	Finisher	2.5	Layer	2.5
Ewe	2.5	Sow	1.0		

Storage/Processing

Can be stored in a mild steel or fibre glass tank.

Alternative Names

Bulk Density

Typical Analysis

Dry Matter	99.5	NCGD	93.0	DUP (@ 5)	-	Avail Lysine	-
Crude Protein	0	NDF	0	DUP (@ 8)	-	Methionine	-
DCP	0	ADF	0	Salt	-	Meth & Cysteine	-
MER	36.0	Starch	0	Ca	-	Tryptophan	-
MEP	37.0	Sugar	0	Total Phos	-	Threonine	-
DE	40.5	Starch + Sugars	0	Av Phos	-	Arginine	-
Crude Fibre	0	FME	1.9	Magnesium	-	PDIA	-
Oil (EE)	98.7	ERDP (@ 2)	-	Potassium	-	PDIN	-
Oil (AH)	98.7	ERDP (@ 5)	-	Sodium	-	PDIE	-
EFA	60.0	ERDP (@ 8)	-	Chloride	-	Met DI	-
Ash	0.5	DUP (@ 2)	-	Total Lysine	-	Lys DI	-

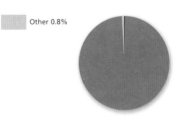

■ Starch 0% ▫ NDF 0% ▨ Other 0.8%

■ Sugars 0% ■ Ash 0.5%

■ Protein 0% ■ Oil 98.7%

Roots, Fruits and By-Products

Introduction

Produced either as a by-product from those grown for human consumption or from plantings to extend forage available.

Origin/Manufacture

UK.

Nutritional Benefit

Swedes and turnips are of similar nutritional value. Swedes provide good levels of energy and sugars in an easily fermented form. They are low in dry matter, but can be used to replace winter forage or concentrate. Swedes have a higher energy value (13.5 MJ/Kg DM) than turnips.

Colour/Texture

Yellow/cream coloured root tubers.

Palatability

Good.

Limits to Usage

Has been reported to taint milk if fed just before milking.

Forage Inclusion % per species

	Inc %		Inc %		Inc %
Calf	0	Creep	0	Chick	0
Dairy	10	Weaner	15	Broiler	0
Beef	20	Grower	15	Breeder	0
Lamb	15	Finisher	0	Layer	0
Ewe	20	Sow	0		

Storage/Processing

Root crops are often stored in clamps during the winter. Dry Matter losses of 10% are not uncommon. Soil contamination should be avoided and the crop should be chopped/sliced for younger animals.

Alternative Names

Bulk Density

Typical Analysis

Dry Matter	11.0	NCGD	82.0	DUP (@ 5)	0.8	Avail Lysine	-
Crude Protein	9.5	NDF	23.3	DUP (@ 8)	1.0	Methionine	-
DCP	7.0	ADF	12.0	Salt	1.0	Meth & Cysteine	-
MER	13.5	Starch	0.1	Ca	0.4	Tryptophan	-
MEP	-	Sugar	59.9	Total Phos	0.25	Threonine	-
DE	-	Starch + Sugars	60.0	Av Phos	-	Arginine	-
Crude Fibre	9.0	FME	13.3	Magnesium	0.1	PDIA	1.5
Oil (EE)	1.0	ERDP (@ 2)	7.5	Potassium	0.5	PDIN	5.8
Oil (AH)	1.2	ERDP (@ 5)	7.1	Sodium	0.2	PDIE	9.5
EFA	-	ERDP (@ 8)	6.9	Chloride	0.2	Met DI	-
Ash	6.0	DUP (@ 2)	0.5	Total Lysine	-	Lys DI	-

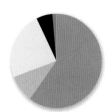

Starch 0.1%	NDF 23.3%	Other 0%
Sugars 59.9%	Ash 6%	
Protein 9.5%	Oil 1.2%	

Roots, Fruits and By-Products	

Introduction

Tubers of Ipomoea batatas (L.) Poir, regardless of their presentation.

Grown as a tuber for human consumption, surplus or down-graded potatoes are dried for export. Usually included in coarse feeds.

Origin/Manufacture

Tropical/temperate countries eg. China.

Nutritional Benefit

High in energy but low in protein. Low in oil (which is relatively unsaturated).
Good source of starch and feeds like normal potatoes.

Colour/Texture

Grey/white. Usually in the form of chips and slices.

Palatability

Good.

Limits to Usage (Anti-Nutritional Factors)

Usually no major anti-nutritive factors, although there have been reports of possible trypsin inhibitors being present. Moulds can grow if stored for long periods or if moisture level is high. Avoid high inclusion rates in ruminants or acidosis will result.

Concentrate Inclusion % per species

	Inc %		Inc %		Inc %
Calf	10	Creep	5	Chick	0
Dairy	20	Weaner	10	Broiler	5
Beef	20	Grower	10	Breeder	10
Lamb	10	Finisher	12.5	Layer	15
Ewe	20	Sow	15		

Storage/Processing

Stores well if moisture is low.

Alternative Names

Bulk Density

Typical Analysis

Dry Matter	87.0	NCDG	93.0	DUP (@ 5)	0.8	Avail Lysine	0.1
Crude Protein	4.0	NDF	7.4	DUP (@ 8)	1.1	Methionine	0.07
DCP	3.0	ADF	1.7	Salt	0.4	Meth & Cysteine	0.1
MER	14.7	Starch	75.0	Ca	0.2	Tryptophan	0.05
MEP	14.8	Sugar	8.5	Total Phos	0.2	Threonine	0.15
DE	16.3	Starch + Sugars	83.5	Av Phos	0.1	Arginine	0.15
Crude Fibre	2.8	FME	14.6	Magnesium	0.1	PDIA	-
Oil (EE)	0.9	ERDP (@ 2)	2.6	Potassium	0.75	PDIN	-
Oil (AH)	1.1	ERDP (@ 5)	2.3	Sodium	0.25	PDIE	-
EFA	0	ERDP (@ 8)	2.0	Chloride	0.25	Met DI	-
Ash	4.0	DUP (@ 2)	0.4	Total Lysine	0.15	Lys DI	-

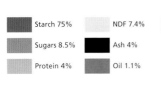

- Starch 75%
- NDF 7.4%
- Other 0%
- Sugars 8.5%
- Ash 4%
- Protein 4%
- Oil 1.1%

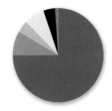

Tapioca

Roots, Fruits and By-Products

Introduction

Roots of Manihot esculenta Crantz, regardless of their presentation.

Tapioca is a tuberous root of a sub-tropical shrub which is processed before feeding to destroy the cyanide present. It is grown for its starch content and the roots are peeled, chopped and dried after harvesting. The material may come as a meal or pellet depending on the processing method. Its usage depends on price and availability of cereal. Availability may also be affected in Europe by import quotas.

Origin/Manufacture

Tropical and Sub Tropical Far East, Africa and S. America.

Nutritional Benefit

Low in protein and oil but high in starch. The protein is heavily made up of non-protein nitrogen (up to 35%). The analysis will also vary depending on the extent of processing. Ideal for ruminants as the starch is slowly degraded and has good energy levels.

Colour/Texture

Muddy white meal/pellet or chips.

Palatability

Can vary depending on cyanide content.

Limits to Usage (Anti-Nutritional Factors)

Linamarin (a glucoside) present releases cyanamide and careful manufacturing is therefore required. Hydrocyanic acid is limited by law and users should consider permitted levels in the feedingstuff regulation.

Concentrate Inclusion % per species

	Inc %		Inc %		Inc %
Calf	5	Creep	0	Chick	5
Dairy	30	Weaner	10	Broiler	10
Beef	30	Grower	15	Breeder	10
Lamb	5	Finisher	30	Layer	15
Ewe	30	Sow	25		

Storage/Processing

Alternative Names

Cassava, Manioc, Manihot.

Bulk Density

Typical Analysis

Dry Matter	87.0	NCGD	80.1	DUP (@ 5)	0.45	Avail Lysine	0.05
Crude Protein	3.0	NDF	15.4	DUP (@ 8)	0.63	Methionine	0.05
DCP	1.1	ADF	6.4	Salt	0.2	Meth & Cysteine	0.07
MER	13.2	Starch	71.0	Ca	0.2	Tryptophan	0.03
MEP	14.9	Sugar	3.0	Total Phos	0.2	Threonine	0.07
DE	15.15	Starch + Sugars	74.0	Av Phos	0.15	Arginine	0.15
Crude Fibre	5.0	FME	13.3	Magnesium	0.15	PDIA	0.8
Oil (EE)	0.6	ERDP (@ 2)	2.1	Potassium	1.1	PDIN	1.9
Oil (AH)	1.4	ERDP (@ 5)	1.8	Sodium	0.05	PDIE	8.5
EFA	-	ERDP (@ 8)	1.6	Chloride	0.15	Met DI	0.05
Ash	6.2	DUP (@ 2)	0.16	Total Lysine	0.1	Lys DI	0.1

- Starch 71%
- Sugars 3%
- Protein 3%
- NDF 15.4%
- Ash 6.2%
- Oil 1.4%
- Other 0%

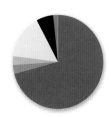

Cereals and By-Products				

Grains of the Triticum X Secale hybrid.

A hybrid seed produced by crossing wheat and rye, which can be grown on marginal land for cereal production.

Origin/Manufacture

Europe.

Nutritional Benefit

It is a naked grain like wheat and of similar nutritive benefit. High energy, moderate protein (although variable) and amino acid content should be checked. Usually contains more lysine and methionine than wheat. Can contain more phosphorous, magnesium and potassium than other grains

Colour/Texture

Naked grain resembling wheat.

Palatability

Good.

Limits to Usage (Anti-Nutritional Factors)

May contain trypsin inhibitors and alkyl products which affect non-ruminants only. Can be low in the amino acid tryptophan. Seeds should be free of ergot to allow feeding.

Concentrate Inclusion % per species

	Inc %		Inc %		Inc %
Calf	20	Creep	20	Chick	10
Dairy	30	Weaner	30	Broiler	20
Beef	35	Grower	35	Breeder	30
Lamb	20	Finisher	40	Layer	35
Ewe	30	Sow	25		

Storage/Processing

Needs processing, eg. rolling, coarse grinding or steam rolling for cattle.

Alternative Names

Bulk Density

720 kg/m³

Typical Analysis

Dry Matter	87.0	NCDG	92.1	DUP (@ 5)	13.7	Avail Lysine	0.37
Crude Protein	14.0	NDF	13.2	DUP (@ 8)	15.0	Methionine	0.25
DCP	11.6	ADF	3.3	Salt	0.05	Meth & Cysteine	0.55
MER	13.5	Starch	63.5	Ca	0.05	Tryptophan	0.15
MEP	14.5	Sugar	4.5	Total Phos	0.6	Threonine	0.45
DE	13.5	Starch + Sugars	68.0	Av Phos	0.3	Arginine	0.60
Crude Fibre	2.5	FME	12.5	Magnesium	0.2	PDIA	3.9
Oil (EE)	2.0	ERDP (@ 2)	11.4	Potassium	0.6	PDIN	8.9
Oil (AH)	2.3	ERDP (@ 5)	10.8	Sodium	0.05	PDIE	10.9
EFA	0.9	ERDP (@ 8)	10.3	Chloride	-	Met DI	0.21
Ash	2.5	DUP (@ 2)	7.0	Total Lysine	0.43	Lys DI	0.77

Starch 63.5%　　NDF 13.2%　　Other 0%

Sugars 4.5%　　Ash 2.5%

Protein 14%　　Oil 2.3%

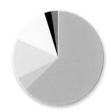

Roots, Fruits and By-Products

Introduction

A mid season (Jul-Aug) or autumn (Sept-Dec) crop, providing a quantity of green feed when grass growth and/or quality is declining. It takes 10 weeks from sowing to feeding.

Origin/Place of Manufacture

Usually grown in more northerly areas of Europe.

Nutritional Benefit

Usually strip grazed, allowing 30-35Kg/head daily. Cows are usually provided with a grass 'run back' area at the same time. 1 acre will last 100 cows for 1 week.

Colour/Texture

Grey/white fleshy tubers.

Palatability

Readily consumed.

Limits to Usage (Anti-Nutritional Factors)

Avoid soil contamination.

Forage Inclusion % per species

	Inc %		Inc %		Inc %
Calf	20.0	Creep	0	Chick	0
Dairy	50.0	Weaner	0	Broiler	0
Beef	50.0	Grower	0	Breeder	0
Lamb	25.0	Finisher	0	Layer	0
Ewe	75.0	Sow	0		

Storage/Processing

Alternative Names

Stubble Turnips.

Bulk Density

Typical Analysis

Dry Matter	10.5	NCDG	82.0	DUP (@ 5)	1.5	Avail Lysine	-
Crude Protein	11.5	NDF	25.4	DUP (@ 8)	1.9	Methionine	-
DCP	8.5	ADF	7.0	Salt	1.0	Meth & Cysteine	-
MER	12.5	Starch	0.1	Ca	0.5	Tryptophan	-
MEP	-	Sugar	55	Total Phos	0.3	Threonine	-
DE	-	Starch + Sugars	55.1	Av Phos	0.25	Arginine	-
Crude Fibre	10.5	FME	12.5	Magnesium	0.1	PDIA	1.1
Oil (EE)	1.0	ERDP (@ 2)	8.2	Potassium	2.0	PDIN	5.9
Oil (AH)	1.5	ERDP (@ 5)	7.7	Sodium	0.5	PDIE	8.7
EFA	-	ERDP (@ 8)	7.3	Chloride	0.7	Met DI	-
Ash	6.5	DUP (@ 2)	1.1	Total Lysine	-	Lys DI	-

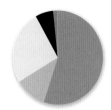

Starch 0.1%	NDF 25.4%	Other 0%
Sugars 55%	Ash 6.5%	
Protein 11.5%	Oil 1.5%	

Miscellaneous	

Introduction

Cattle and sheep have the ability through their rumen to utilize urea/non protein nitrogen (NPN) to manufacture microbial protein. Only feed grade urea should be used as fertilizer grade may contain heavy metals.

Origin/Place of Manufacture

Widely manufactured in Europe, USA and S. America.

Nutritional Benefit

Urea is approximately 50% nitrogen and is the most concentrated source available. As protein is N x 6.25, urea is equivalent to 295% Crude Protein. Care should be taken to supply urea in a safe form as it is toxic in large amounts and can produce ammonia toxicity. Urea must be fed with a readily available energy source to ensure rumen capture.

Urea for Grain Treatment: Feed grade urea can be used to preserve moist grain (20-30% moisture). Moisture below 20% will result in inadequate preservation as there is insufficient moisture to hydrolize the urea fully.

Grain Moisture	kg/Urea/tonne Grain
20 - 22	25
23 - 25	30
26 - 28	35
> 28	40

The grain should be well mixed with the urea and an addition at 20-40 kg of water added in the mixture. When mixed, it should be stored in a clean area covered with a plastic sheet, to retain ammonia vapour released by the process at a height of 1 metre, to prevent overheating.

Colour/Texture

White/grey granules.

Palatability

Poor.

Limits to Usage

Typical feed rates for milking cows range from 50-150 grams per head per day in a well balanced complete diet. Care must be taken on overall levels of non-protein nitrogen fed to ruminants as ammonia toxicity can easily result.

Concentrate Inclusion % per species

	Inc %		Inc %		Inc %
Calf	0	Creep	0	Chick	0
Dairy	0.015	Weaner	0	Broiler	0
Beef	0.015	Grower	0	Breeder	0
Lamb	0	Finisher	0	Layer	0
Ewe	0.01	Sow	0		

Storage/Processing

Alternative Names

Bulk Density

550 - 600 Kg/m³

Typical Analysis

Dry Matter	99.5	NCDG	0	DUP (@ 8)	0	Meth & Cysteine	0
Crude Protein	280.0	NDF	0	Salt	0	Tryptophan	0
DCP	236.0	ADF	0	Ca	0	Threonine	0
MER	0	Starch	0	Total Phos	0	Arginine	0
MEP	0	Sugar	0	Av Phos	0	PDIA	0
DE	0	Starch + Sugars	0	Magnesium	0	PDIN	68.0
Crude Fibre	0	FME	0	Potassium	0	PDIE	0
Oil (EE)	0	ERDP (@ 2)	230.0	Sodium	0	Met DI	0
Oil (AH)	0	ERDP (@ 5)	230.0	Chloride	0	Lys DI	0
EFA	0	ERDP (@ 8)	230.0	Total Lysine	0		
Linoleic	0	DUP (@ 2)	0	Avail Lysine	0		
Ash	1.0	DUP (@ 5)	0	Methionine	0		

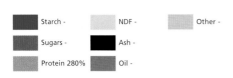

Starch - NDF - Other -

Sugars - Ash -

Protein 280% Oil -

Cereals and By-Products

Introduction

Grains of Tritcum aestivum L., Tritcum durum Desf. and other cultivars of wheat.

Wheat is classified into three types - hard, soft, durum. It is used in bread, pasta, brewing and starch manufacture, with soft varieties generally used as animal feed. The naked grain has a prominent crease and is oval in shape. Wheat can now be fed at higher levels in compound feeds due to the inclusion of enzymes which reduce stickiness. Feed wheat in the UK has been grown for bread or biscuit making qualities but failed to attain the grade, hence unavailability for feed. The most common use of wheat is to make bread which requires hard wheats, with high proteins, high Hagberg falling numbers so it can make a stiff dough.

Origin/Manufacture

The most common cereal in Europe and other temperate countries.

Nutritional Benefit

Very high in energy with average protein (13%). High in starch (64%), low in fibre (3% as a naked grain), but tends to be low in vitamins especially biotin. Vitamin E is reduced when grain is stored moist with preservatives. It is useful for increasing milk protein yields and for the promotion of growth. Approximately 10% of the starch is rumen unfermented.

Colour/Texture

Pale brown oval grain.

Palatability

Good.

Limits to Usage (Anti-Nutritional Factors)

Contains high levels of gluten which, if exessively ground, can result in a sticky dough, reducing digestion. The readily fermentable carbohydrate present can cause acidosis when fed at high levels to ruminants.

Concentrate Inclusion % per species

	Inc %		Inc %		Inc %
Calf	25	Creep	60	Chick	50
Dairy	40	Weaner	55	Broiler	60
Beef	40	Grower	50	Breeder	65
Lamb	25	Finisher	50	Layer	60
Ewe	35	Sow	50		

Storage/Processing

Can be crushed and rolled or coarsely ground (2mm for Pigs, 4mm+ for Poultry). Sheep can eat whole grains. Wheat will improve pellet quality (10% minimum).

Alternative Names

Bulk Density

700 - 770 Kg/m³

Typical Analysis

Dry Matter	86.0	NCDG	93.5	DUP (@ 5)	1.1	Avail Lysine	0.3
Crude Protein	13.0	NDF	12.0	DUP (@ 8)	1.4	Methionine	0.21
DCP	10.0	ADF	2.6	Salt	0.15	Meth & Cysteine	0.45
MER	13.8	Starch	67.0	Ca	0.06	Tryptophan	0.15
MEP	15.1	Sugar	4.0	Total Phos	0.35	Threonine	0.4
DE	16.0	Starch + Sugars	71.0	Av Phos	0.15	Arginine	0.6
Crude Fibre	3.0	FME	12.8	Magnesium	0.15	PDIA	2.7
Oil (EE)	1.8	ERDP (@ 2)	10.7	Potassium	0.5	PDIN	8.2
Oil (AH)	2.0	ERDP (@ 5)	10.3	Sodium	0.05	PDIE	10.2
EFA	1.5	ERDP (@ 8)	9.9	Chloride	0.01	Met DI	0.19
Ash	2.0	DUP (@ 2)	0.7	Total Lysine	0.35	Lys DI	0.2

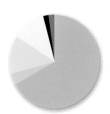

Starch 67%	NDF 12%	Other 0%	
Sugars 4%	Ash 2%		
Protein 13%	Oil 2%		

Introduction

By-product of flour manufacture, obtained from screened grains of wheat or dehusked spelt. It consists principally of fragments of the outer skins and of particles of grain from which the greater part of the endosperm has been removed.

This is the husk from grains sold as a by-product of flour manufacture with a low bulk density. It consists of the pericarp and testa or coarse bran.

Origin/Manufacture

UK.

Nutritional Benefit

High in phosphorus, magnesium and low in calcium. Energy value and starch levels are low.

Colour/Texture

Light tan coloured fluffy fibrous meal/pellet.

Palatability

Average.

Limits to Usage

No anti-nutritive factors, although the physical nature may limit inclusion rates.

Concentrate Inclusion % per species

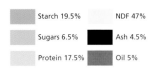

	Inc %		Inc %		Inc %
Calf	10	Creep	0	Chick	0
Dairy	20	Weaner	0	Broiler	5
Beef	25	Grower	5	Breeder	5
Lamb	5	Finisher	10	Layer	5
Ewe	20	Sow	25		

Storage/Processing

Alternative Names

Bulk Density

300 - 400 Kg/m³

Typical Analysis

Dry Matter	86.0	NCGD	71.0	DUP (@ 5)	2.0	Avail Lysine	-
Crude Protein	17.5	NDF	47.0	DUP (@ 8)	2.7	Methionine	0.3
DCP	15.2	ADF	12.8	Salt	0.15	Meth & Cysteine	0.6
MER	11.0	Starch	19.5	Ca	0.15	Tryptophan	0.25
MEP	7.5	Sugar	6.5	Total Phos	1.3	Threonine	0.6
DE	11.5	Starch + Sugars	26.0	Av Phos	0.5	Arginine	1.2
Crude Fibre	12.0	FME	9.3	Magnesium	0.55	PDIA	4.3
Oil (EE)	4.0	ERDP (@ 2)	14.6	Potassium	1.3	PDIN	11.8
Oil (AH)	5.0	ERDP (@ 5)	13.6	Sodium	0.05	PDIE	10.0
EFA	2.5	ERDP (@ 8)	12.6	Chloride	0.1	Met DI	0.2
Ash	4.5	DUP (@ 2)	1.0	Total Lysine	0.7	Lys DI	0.68

Starch 19.5% NDF 47% Other 0%

Sugars 6.5% Ash 4.5%

Protein 17.5% Oil 5%

Cereals and By-Products

Introduction

This is whole feed wheat grain treated with sodium hydroxide (caustic) to rupture the seed coat and make the whole grain digestible. It has grown in popularity and can be processed in a feeder wagon and is seen as an alternative to grinding or rolling.
Treatment Process: Add grain to feeder wagon. Add caustic (3%) and mix for 5 mins. Add water to correct to 65-70% dry matter (in mixture). Mix for 10 mins and leave to stand in a heap for 10 hours (it will heat up). Level out and leave to cool. Feed after 3-4 days. The NaOH converts to Sodium Bicarbonate (NaHCO3) and impregnates the grains surface.

Origin/Place of Manufacture

UK, Ireland, Germany, Denmark, Sweden, France, Holland, USA.

Nutritional Benefit

High in starch of which 70% is rumen unfermented. The sodium bicarbonate coating is claimed to act as a natural buffer to the acids produced in ruminants when this processed wheat is fed. Whole grain reduces the rate of acid load in the rumen due to the low surface area.

Colour/Texture

Dark brown/golden. May darken the longer it is stored. A white 'bloom' may develop on the surface, this is sodium bicarbonate formed during the processing.

Palatability

Good in a total mixed ration.

Limits to Usage (Anti-Nutritional Factors)

High in sodium and needs a low sodium mineral balancer. Vitamin E is also reduced where grains are stored moist.

Forage Inclusion % per species

	Inc %		Inc %		Inc %
Calf	10	Creep	0	Chick	0
Dairy	25	Weaner	0	Broiler	0
Beef	25	Grower	0	Breeder	0
Lamb	0	Finisher	0	Layer	0
Ewe	20	Sow	0		

Storage/Processing

Material made at 65-70% will store well at ambient temperature inside or outside (if covered) for up to 3 weeks. For longer term storage, less water should be added with a target dry matter of 80%. Such material under clean dry conditions will keep for up to 6 months. Sodium Hydroxide is a hazardous chemical and should be handled with care, adhering to safety instructions of supplier.

Alternative Names

Soda Wheat.

Bulk Density

Caustic Wheat, Soda Grain.

Typical Analysis

Dry Matter	75.0	NCGD	92.5	DUP (@ 5)	0.7	Avail Lysine	-
Crude Protein	12.5	NDF	11.0	DUP (@ 8)	0.9	Methionine	0.2
DCP	10.0	ADF	2.5	Salt	0.15	Meth & Cysteine	0.4
MER	13.5	Starch	66.0	Ca	0.05	Tryptophan	0.1
MEP	-	Sugar	4.0	Total Phos	0.4	Threonine	0.4
DE	-	Starch + Sugars	70.0	Av Phos	-	Arginine	0.6
Crude Fibre	3.0	FME	13.0	Magnesium	0.1	PDIA	2.7
Oil (EE)	2.0	ERDP (@ 2)	10.5	Potassium	0.5	PDIN	8.4
Oil (AH)	2.0	ERDP (@ 5)	10.4	Sodium	3.5	PDIE	10.5
EFA	-	ERDP (@ 8)	10.2	Chloride	0.1	Met DI	-
Ash	4.5	DUP (@ 2)	0.5	Total Lysine	0.3	Lys DI	-

Starch 66%	NDF 11%	Other 0%
Sugars 4%	Ash 4.5%	
Protein 12.5%	Oil 2%	

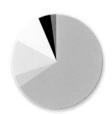

CONTEXT

Cereals and by-products

Introduction

By-product of flour manufacture from screened grains of wheat or dehusked spelt. It consists principally of fragments of the outer skins and of particles of grain from which less of the endosperm has been removed than in wheat bran.

This is the 'family' name given to all the offals from screening and de-husking of wheat from the flour milling industry including wheat bran, endosperm and other screenings. For every tonne of wheat processed, approximately 20 kg of wheat feed is produced.

Origin/Place of Manufacture

Mainly UK.

Nutritional Benefit

Highly digestible, but quality varies between production plants. Ideal for ruminants and pig feed and also for horse feed. A good source of phosphorus, but low in calcium, sodium and vitamins.

Colour/Texture

Light brown, fibrous meal.

Palatability

Good.

Limits to Usage (Anti-Nutritional Factors)

Concentrate Inclusion % per species

	Inc %		Inc %		Inc %
Calf	20	Creep	5	Chick	0
Dairy	30	Weaner	10	Broiler	5
Beef	45	Grower	25	Breeder	15
Lamb	20	Finisher	25	Layer	15
Ewe	30	Sow	20		

Storage/Processing

Goes mouldy quickly and will develop strange flavours.

Alternative Names

Thirds, Midds, Sharps or Wheatings.

Bulk Density

Meal 350 Kg/m³ **Pellets** 560 Kg/m³

Typical Analysis

Dry Matter	86.0	NCDG	74.0	DUP (@ 5)	4.1	Avail Lysine	0.5
Crude Protein	17.5	NDF	40.0	DUP (@ 8)	5.2	Methionine	0.4
DCP	15.3	ADF	10.2	Salt	0.15	Meth & Cysteine	0.7
MER	11.5	Starch	27.5	Ca	0.15	Tryptophan	0.25
MEP	10.5	Sugar	6.5	Total Phos	1.2	Threonine	0.7
DE	11.5	Starch + Sugars	34.0	Av Phos	0.8	Arginine	1.1
Crude Fibre	9.0	FME	10.0	Magnesium	0.5	PDIA	4.5
Oil (EE)	3.5	ERDP (@ 2)	13.6	Potassium	1.5	PDIN	12.3
Oil (AH)	4.0	ERDP (@ 5)	12.0	Sodium	0.05	PDIE	10.8
EFA	2.3	ERDP (@ 8)	10.8	Chloride	0.07	Met DI	0.21
Ash	4.5	DUP (@ 2)	2.6	Total Lysine	0.8	Lys DI	0.72

Starch 27.5%   NDF 40%   Other 0%

Sugars 6.5%   Ash 4.5%

Protein 17.5%   Oil 4%

Miscellaneous	

Introduction
When milk is treated with rennet in the process of cheese making, casein is precipitated and takes with it most of the fat and almost half the calcium and phosphorus. The liquid residue is whey.

Origin/Manufacture
UK.

Nutritional Benefit
It is a poorer source of energy, fat, soluble vitamins, calcium and phosphorus than milk. It is often fed in wet feeding systems to pigs.

Colour/Texture
Yellow/white liqour.

Palatability
Good when fresh.

Limits to Usage (Anti-Nutritional Factors)
Low dry matter and high salt content.

Concentrate Inclusion % per species

	Inc%		Inc%		Inc%
Calf	5	Creep	0	Chick	0
Dairy	10	Weaner	10	Broiler	0
Beef	10	Grower	20	Breeder	0
Lamb	5	Finisher	20	Layer	0
Ewe	10	Sow	10		

Storage/Processing

Alternative Names

Bulk Density

Typical Analysis

Dry Matter	47.0	NCGD	0	DUP (@ 5)	-	Avail Lysine	-
Crude Protein	22.0	NDF	0	DUP (@ 8)	-	Methionine	-
DCP	17.0	ADF	0	Salt	12.0	Meth & Cysteine	-
MER	13.0	Starch	0	Ca	0.8	Tryptophan	-
MEP	-	Sugar	12.0	Total Phos	1.2	Threonine	-
DE	10.8	Starch + Sugars	12.0	Av Phos	-	Arginine	-
Crude Fibre	0	FME	-	Magnesium	0.2	PDIA	-
Oil (EE)	0	ERDP (@ 2)	-	Potassium	3.3	PDIN	-
Oil (AH)	0	ERDP (@ 5)	-	Sodium	1.3	PDIE	-
EFA	0	ERDP (@ 8)	-	Chloride	3.2	Met DI	-
Ash	10	DUP (@ 2)	-	Total Lysine	-	Lys DI	-

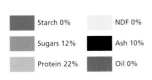

- Starch 0%
- NDF 0%
- Other 56%
- Sugars 12%
- Ash 10%
- Protein 22%
- Oil 0%

Cereals and By-Products

Introduction

Whole crop cereals provide a means of producing a self dessicating high energy forage and can be produced from either autumn (preferable) or spring grown cereals. The crop is either fermented at a lower Dry Matter (35%) or feed grade urea is used to preserve it when harvested at high Dry Matters (50%). Well fermented whole crop wheat can have feed values close to maize silage. Poorer whole crops are indifferent forages. For fermented material, harvest at the cheesey dough stage and clamp quickly. For urea treatment, cut at 55-60% Dry Matter and add urea at 20-30 kg per tonne of fresh forage. Wheat is preferable to barley, which forms a tough grain coat once the crop exceeds 35% Dry Matter, resulting in poor digestibility.

Origin/Manufacture

UK, Ireland, S. America, USA, Australia, Denmark.

Nutritional Benefit

Urea treated wholecrop has good Dry Matter (50-60%) and is high in protein (mainly ammonia), but needs a soluble carbohydrate, eg. molasses, to capture the free ammonia by the rumen bugs. It has good levels of starch. It is alkaline and balances the acidity of grass silage. Up to 15 tonnes of fresh weight can be produced per acre or 15 tonnes of dry matter/hectare. Ideal as a buffer feed and can be fed as 40% of the forage dry matter intake or a maximum of 5kg dry matter per head/per day. Urea treatment results in high crude protein and amino contents so care is essential to ensure the degradeable content of the total ration is not excessive.

Colour/Texture

Yellow/brown, fibrous forage.

Palatability

Fermented - good. Urea treated - average.

Limits to Usage (Anti-Nutritional Factors)

Its high ammonia content, when urea treated, needs attention to avoid excess degradable protein in the diet. High straw fraction limits inclusions for very productive stock.

Forage Inclusion % per species

	Inc %		Inc %		Inc %
Calf	20	Creep	0	Chick	0
Dairy	40	Weaner	0	Broiler	0
Beef	40	Grower	0	Breeder	0
Lamb	20	Finisher	0	Layer	0
Ewe	34	Sow	0		

Storage/Processing

A fine chop is suggested and an air-tight seal is essential. The urea breaks down to ammonia,which spreads through the crop, breaking down fibrous material. It stores well as mould and bacteria are inhibited by the alkalinity. Store in a narrow clamp to minimise face deterioration.

Alternative Names

Bulk Density

544 - 672 Kg/m³

Whole Crop Fermented

Typical Analysis

Dry Matter	40.0	NCGD	54.0	DUP (@ 5)	1.5	Total Lysine	-
Crude Protein	9.5	NDF	54.0	DUP (@ 8)	1.8	Methionine	-
DCP	7.0	ADF	35.0	Salt	0.05	Meth & Cysteine	-
MER	10.5	Starch	25.0	Ca	0.2	Tryptophan	-
MEP	-	Sugar	1.0	Total Phos	0.25	Threonine	-
DE	-	Starch + Sugars	26.0	Av Phos	-	Arginine	-
Crude Fibre	23.0	FME	8.5	Magnesium	0.1	PDIA	2.0
Oil (EE)	2.5	ERDP (@ 2)	6.0	Potassium	1.5	PDIN	5.5
Oil (AH)	3.0	ERDP (@ 5)	5.5	Sodium	0.02	PDIE	6.3
EFA	-	ERDP (@ 8)	5.2	Chloride	-	Met DI	-
Ash	7.5	DUP (@ 2)	1.2	Avail Lysine	-	Lys DI	-

Whole Crop Urea

Typical Analysis

Dry Matter	55.0	NCGD	57.0	DUP (@ 8)	3.5	Meth & Cysteine	-
Crude Protein	25.0	NDF	54.0	Salt	0.05	Tryptophan	-
DCP	12.0	ADF	31.0	Ca	0.2	Threonine	-
MER	10.0	Starch	25.0	Total Phos	0.25	Arginine	-
MEP	-	Sugar	2.5	Av Phos	-	PDIA	3.3
DE	-	Starch + Sugars	27.5	Magnesium	0.1	PDIN	9.3
Crude Fibre	23.0	FME	8.5	Potassium	1.5	PDIE	8.0
Oil (EE)	2.0	ERDP (@ 2)	10.1	Sodium	0.05	Met DI	-
Oil (AH)	2.5	ERDP (@ 5)	9.1	Chloride	-	Lys DI	-
EFA	-	ERDP (@ 8)	3.1	Total Lysine	-		
Linoleic	-	DUP (@ 2)	2.5	Avail Lysine	-		
Ash	7.5	DUP (@ 5)	3.0	Methionine	-		

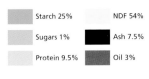

Starch 25%   NDF 54%   Other 0%

Sugars 1%   Ash 7.5%

Protein 9.5%   Oil 3%

Notes

Notes